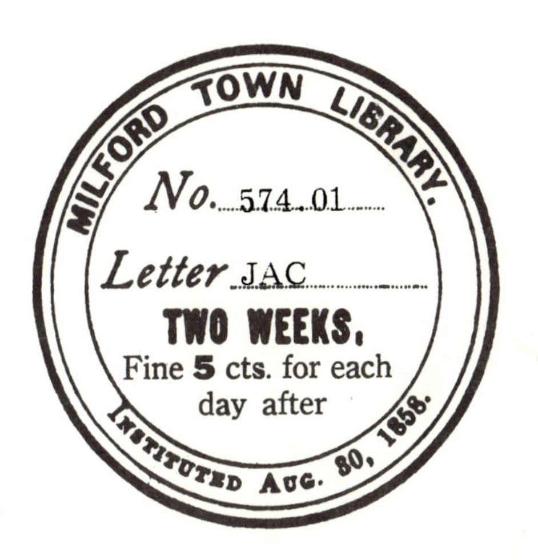
MILFORD TOWN LIBRARY.
No. 574.01
Letter JAC
TWO WEEKS.
Fine 5 cts. for each day after
INSTITUTED AUG. 30, 1858.

ENDANGERED BY SCIENCE?

Endangered by Science?

Albert Jacquard

Translated by Margaret M. Moriarty

Columbia University Press
NEW YORK
1985

Columbia University Press
New York Guildford, Surrey

Printed in the United States of America

Library of Congress Cataloging in Publication Data

Jacquard, Albert.
Endangered by science?

Translation of: Au péril de la science?
Includes index.
1. Human population genetics—Philosophy. 2. Human population genetics—Social aspects. I. Title.
QH455.J26713 1985 574'.01 85-5690
ISBN 0-231-05694-X (alk. paper)

Clothbound editions of Columbia University Press books are Smyth-sewn and printed on permanent and durable acid-free paper.

Contents

ENDANGERED BY SCIENCE?

Introduction

During the latter half of the nineteenth century, it became customary for Science to be written with a capital S and to have tremendous faith pinned on it. Rejecting obscurantism, breaking the spell of old myths, removing ancient fears, renouncing cowardly submissions, at last examining the universe with a lucid and open mind, dominating it by knowing it better, acting on it, transforming it, subjugating it, shaping the future of humanity—all this was going to be possible thanks to scientific progress.

Beyond the sonorous verbiage of official inaugurations or prize-giving ceremonies, a genuine faith had developed, profoundly changing each person's attitude toward his destiny: the future was no longer anticipated with fear but with hope.

A century has passed. The fruits are even more numerous than predicted, but they are bitter. The world has indeed been transformed, alas! Man has taken possession of the planet to the point where it is unrecognizable. A diffuse anxiety has spread; we hear predictions that are more sinister than any in previous human experience, and, nevertheless, what has already been done is merely a sample of what could be done, of what, perhaps, is being prepared. What scientists have on display is nothing compared to what they have in stock. Henceforth, humanity lives under a permanent threat, with little hope of ever removing it. A handful of men could decide to blot out all life from our planet in a few seconds. We all know this, but we try not to think about it, for fear of being obliged to think about it constantly. Will we have to live with this obsession until the end of time?

While science still inspires some people with hope, it has become a source of fear to many. An attitude of rejection has appeared, and has gradually spread. This rejection is sometimes presented as the only means of avoiding ultimate catastrophe, and it is easily justified by the excesses to which scientific efficacity have led. Those whose imaginations do not stretch to visions of nuclear holocaust can simply observe the deterioration of the countryside: even the wheat fields, once vibrant with colorful poppies and singing birds, have become, for the sake of improving the yield, huge and sinister disinfected "concentration camps" (E. Morin) for vegetables classified by species.

These achievements, gifts of science, do they not suffice to condemn it outright while there may still be time?

Some scientists who are themselves genuinely upset by the foreseeable consequences of the work to which they are contributing set the tone; while often seeming offhand and sometimes coldly ironic, they talk unreservedly about their anguish, but nonetheless continue with their research. Passengers along with their contemporaries on the same blind train, they continue to stoke the engine up to the hilt while pulling the alarm signal and waiting for others to pull the brake.

Their reluctance is understandable, because the outcome is not only negative:

—Hunger, disease, and death have receded. To illustrate this point, it will suffice to mention one wonderful achievement, which no one would have dared hope for as little as twenty years ago, and which now seems assured: the smallpox virus which, every year, used to afflict, blind, or kill millions of people, has been completely eradicated; it is now found in only seven laboratories, where it is carefully confined within a few glass phials. This event, more decisive for the history of humanity than many of the battles recorded in our history books, can be precisely dated: it was on October 26, 1977, that the last case of smallpox was found, in Somalia. Does not "1977" one day deserve to take the place of "Marignano 1515" or "1914–1918" in our memories?

—The ancient curse "You will earn your bread by the sweat of your brow" is beginning to become obsolete. For an increasing number of people, life is no longer just a perpetual quest for the means of survival; because of the technical inventions which have followed the advance of knowledge, our ability to create wealth has developed to a point where the privilege of leisure could, without undue difficulty, be extended much more widely.

One could endlessly prolong this list of the benefits and misdeeds of science, in a vain search for a final assessment. It is, however, necessary to pursue this line of thought: science is not an autonomous tree, developing according to its own laws, and from which we can passively gather the fruit; it is a collective enterprise, our own, and it is up to us to determine its direction. Both the proscientific pronouncements of the late nineteenth century and the antiscientific ones of the late twentieth are equally futile: what is important is to understand the process with which we are dealing and in which we are participating. And, first of all, we need to ask ourselves what exactly do we mean by the word "science."

This kind of questioning can be meaningful only if we are careful to go beyond generalities, and enter into the day-to-day reality of those who are actually engaged in the scientific enterprise. To do this, it is necessary to focus our analysis on one specific field of research; throughout this book, the examples will be taken from one discipline, the only one in which the author has some experience, genetics, and, even more specifically, population genetics. But the questions raised by the development of this discipline lead to broader questions that embrace all scientific domains.

1.
Science and Us

1. Science, a Human Achievement

ARGUMENTS BASED ON AUTHORITY AND SCIENTIFIC ARGUMENTS

When we make a statement, we feel the need to justify it and bolster it with argument.

In many cases, we content ourselves, let's face it, with an argument from authority. This is true, because so and so said so: "Aristotle said so," "the Church Fathers tell us." This argument does of course give us a feeling of intellectual comfort by putting a few prestigious names on our side, but it could hardly convince someone who was not already convinced. The development of a scientific attitude consisted, notably, in a refusal of this type of argument: it involves proving by rational argument and linking one's statement, through logical deduction, to a body of previously accepted facts or doctrines.

This procedure is, to be sure, much more cumbersome and laborious, and it is tempting to use its end product without bothering to go through all the preliminary stages. Suddenly, one is back, beneath a seemingly scientific approach, to a simple argument from authority. The only difference is the replacement of the philosophers of antiquity or the Church Fathers, constantly invoked in the Middle Ages, by winners of the Nobel Prize or the Fields Medal.

Certain psychiatrists dabbling in genetics and certain clubs of young technocrats trying to promote an elitist society lace their statements with incantations to "modern science" or to "recent discoveries in biology." This is particularly clear in the case of the current revival of racism in all its forms. It is always in the name of Science that attempts at defining and hierarchizing the various human groups are made. However, it is easy to ascertain that the content of the "recent discoveries" invoked, and their links with the arguments being presented, are seldom specified.

In our society, the prestige of scientific method is such that most people feel obliged to agree with any statement that is labeled "scientific." This is because, in the mind of those who use this term in this way, science is, by definition, that which cannot be questioned. And this attitude is quite widespread.

As its etymology suggests, science is seen by most people as a body of knowledge, facts, and certainties. In the eyes of most of our contemporaries, it is a magnificent edifice being gradually constructed by researchers. Its completion seems to be still a remote prospect, but it is progressing rapidly. The aim of this enterprise is to understand the universe and change it; to do this, its parameters must be defined and measured, the meaning of the facts observed must be deciphered, the laws governing the relationships between the various constituents of this universe must be deduced and shown to coincide with experimental evidence. Enriched by the achievements of his predecessors, the scientist advances; just as explorers narrowed and then eliminated the *terra incognita,* the white blotches on the globe, so now are scientists narrowing the dark areas of knowledge.

Within this perspective, the criterion for the accuracy and authenticity of this science is its efficacy: with his more accurate understanding of natural processes, man can take action, subjugate the elements, inhabit (the only species capable of so doing) the entire surface of the earth, conquer sickness, sometimes death, even escape his planet and explore first the Moon and eventually the other planets. It is because of these successes that science is revered and that scientists are listened to and used as guarantors.

This description, however, is quite different from science as it is seen from the inside, and as it is practiced. Science is first and foremost a practice, that is, an activity requiring adherence to certain rules.

SCIENTIFIC PRACTICE

This practice involves producing discourse about the universe (atomic particles or galaxies, viruses or human societies). In order to be deemed worthy of contributing to science, this discourse must satisfy certain requirements that are more or less explicitly accepted by the scientific community.

The first rule is that only words whose meaning has been clearly specified can be used. This seems to be a minimal requirement, one that is normally applicable to any kind of discourse, where there is no question of science. Even a cursory look at the ocean of statements that beseiges us each day shows that unfortunately this is not so. How many statements there are with which it is impossible either to agree or to disagree, so obscure is their meaning! For instance, to stay on the question of racism, consider most pronouncements on the "superiority" or "inferiority" of a particular "race" or on the "genetic determinism of intelligence." Rarely do we find a precise definition of any of these words.

Another rule concerns the possibility of imagining a way of refuting the statements contained in this discourse. This is the logician Popper's famous "refutation principle"; he considers this principle to be the criterion that distinguishes scientific statements from ideological ones. It is perfectly acceptable to proclaim that "Whites must dominate blacks" (or the opposite), "Long live the king!" (or "Up the republic!"), but this has nothing to do with science because these statements cannot be subjected to verification.

The primary objective of scientific enterprise is not, contrary to a widely held belief, effective intervention in the world around us, but the most coherent possible representation of it; it is above all a quest for lucidity.

To illustrate the split between science as it is seen by our contemporaries (and used by some) and science as it is experienced (or should be) by those who actually do it, let us imagine putting the following question to the average person: "What were the most important scientific achievements in the twentieth century?"

Many would immediately mention the invention of new procedures or substances such as penicillin, which made it possible to save so many human lives, or the use of nuclear energy, which made it possible to eliminate so many. These discoveries gave man the *means of intervening* (for worse as well as for better, to be sure, but that is thought to be the responsibility of politicians and not of scientists).

Others would cite the theory of relativity or the double helix of DNA. Thanks to Einstein, we have a better understanding of the space-time in which we are evolving; thanks to Crick and Watson, we now see how the making of proteins and the reproduction of biological information, necessary to the maintenance and transmission of life, are ensured. These discoveries gave us the *means of understanding*.

However, many scientists would respond to this question in quite a different manner. They would focus not on the fact that it affords us a better understanding of the real world and therefore an improved capacity for changing it, but on the way it enhances our ability to formulate our questions about it. A scientist's attitude is entirely different from that of a landowner who, on acquiring a new piece of land, appraises its soil, imagines the best means of cultivating it, and dreams of the harvest to come; on the contrary, when new terrain comes his way, he hurries to its boundaries and is obsessed by just one question: what is hidden behind the walls that surround it, and how can he get over them? The most decisive scientific advances are those which give us the means of asking better questions.

It so happens that our century has been, in this regard, especially fertile. However, the newest and most innovative concepts have been camouflaged by the plethora of technological successes, which often pass for science. The wonder and amazement caused by the new powers which man has acquired, thanks to his new expertise, prevent us from becoming fully aware of conceptual changes which occurred simultanously. There can thus be a considerable time lag between the development of a concept by scientists and its diffusion among the general public.

One particularly clear case of this is that of our understanding of so-called "sexual" living organisms. Not until the beginning of the twentieth century was the discovery originally made by Mendel in 1865 finally understood: the individual, in spite of appearing to be an indivisible unit, is in reality a being with dual controls, each of whose fundamental traits depends not on one, but on two factors. This "duplicity" is so contrary to common sense that it could only be accepted by the scientific community after half a century, and it is still not truly accepted by the general public. This profound incomprehension of the central process of the transmission of life, the conception of a child, is clearly illustrated by the terms used in sex education manuals to explain it to children: "So that you could be born, your daddy put a seed into your mommy's tummy"; this is said with the best of intentions and draws general agreement; however, it is completely wrong, because this way of putting it denies the symmetry of the two parents' roles by assigning to one the active role of the sower and to the other the passive role of the soil. The actual mechanism by which each parent transmits half of the biological information which he had himself received is beyond our imagination, and the very words which we use distort the truth.

Our comprehension of the living world has progressed within a general conceptual framework which has been profoundly altered in the course of this century. The thinking of biologists is permeated by ideas which, unknown to them, seem very far removed from their domain. Having surfaced in one discipline, they gradually spread to all other domains and transform the very way in which problems are tackled, but this diffusion is sometimes strangely slow. Consider, for example, two concepts which were introduced fifty years ago and which have become essential components of the scientific approach: "Uncertainty" and "undecidability."

UNCERTAINTY

The development of modern physics necessitated some fundamental rethinking. The behavior of "elementary particles" can no longer be explained using Descartes' and Newton's terms; new concepts have to be hammered out which require a new attitude to the real world, one which often involves realizing that in giving certain explanations, in claiming to bend this behavior to our logic, we are quite simply talking a lot of hot air.

To describe this behavior, classical physics relies on measures which, in the case of a material particle, specify its mass, its position, and its speed. In 1927, the physicist Heisenberg published a result based on an argument that was developed using the concepts of quantitative physics and which, expressed in terms of the vocabulary of classical physics, is known as "uncertainty equations." According to this interpretation, the degree of precision with which we can simultaneously estimate the position and speed of a material element is necessarily below a certain threshold. The better we know the position, the less well we know the speed, and vice versa. In reality, any measure requires an observation and therefore a distortion caused by the interaction between the thing observed and the instrument being used to observe it. This inevitable distortion raises an insurmountable barrier to the precision of our knowledge of the real world.

This finding obviously destroys the old dream of the perfect predictability of the universe, which was expressed, especially in the eighteenth century, by the apologue of the "demon of Laplace." The state of each particle, Laplace tells us, is defined by a certain number of parameters; it is subject in an absolute way to certain laws which it cannot escape—for instance, the law of universal gravity, which specifies the intensity of the attraction exerted by other bodies on this particle as a function of its mass, of their masses, and the distances between it and them. A being (a "demon") capable, at a given second, of knowing all the parameters of all the particles in the universe, and acquainted with all of its laws, would be able to predict all the changes, to describe the state of the universe at the next second and, step by step, all the states to come; similarly, he could reconstitute the state it was in a second before that and, gradually, the entire history of the universe since its origin. In this conception, knowledge of the present implies knowledge of the past and of the future, time is abolished, everything is determined.

Heisenberg's result has often been interpreted as proof of man's definitive incapacity: never will he be able to obtain the absolute knowledge of the present required by the calculations of Laplace's "demon"; being part of the universe, he cannot embrace it with a glance from outside of it; he cannot gather information without this quest for information itself becoming a source of distortion, changing the object that he wishes to know. But some argue that this limitation to human power does not for all that mean that this universe is indeterminable: reality, at a given second *t*, exists, even if we do not have complete access to it; it is a necessary product of the reality which preceded it, it determines unambiguously the reality which is to follow, because, as Leibniz said, "The present is pregnant with the future" or, as Einstein said, "God does not play dice." In this interpretation of "uncertainty," the essence of the Laplacian vision of a perfectly autodetermined world is preserved; it is only the human capacity for acceding to this reality that is questioned, but not the very existence of this reality and the inexorable unfolding of the strictly inevitable process of its evolution.

However, to certain "quantitative" physicists, the presentation of Heisenberg's result in terms of uncertainties betrays its real significance. The question raised is, in their view, much more radical. What is at stake is the possibility, independently of the limits of human faculties, of describing nature from a single point of view; when we ask questions of it, we inevitably use a particular language to do so; the reply is limited by this very language; to quote Prigogine and Stengers, "There is no viewpoint from which the totality of the real world would be visible simultanously. . . . The richness of the real world exceeds every language, every logical structure, every conceptual framework."[1]

We have become accustomed to thinking of elementary particles as though they were objects like any others, except that they are much smaller; we therefore think it natural to describe them in terms of mass, position, and speed. Similarly, we have become accustomed to thinking of light as being a kind of wave, which we characterize, for example, by its frequency. Quantitative physics leads us to consider objects which are neither waves nor particles, but which can be described, depending on the point of view adopted, by means of the measurements that characterize both of them.

The danger of purely verbal explanations camouflaging, beneath misleading images, the irreducible complexity of reality is illustrated by the famous model of the atom as a miniature solar system: electrons rotate around the nucleus, like the planets around the sun. In fact, the electron in its orbit is assumed to emit no energy, a necessary hypothesis for this orbit to be stable; strictly speaking, its position is therefore unknowable, since, to pinpoint it, one would need to see this electron, that is, to pick up a photon emitted by it. Now the electron can emit a photon and supply information about itself only by changing orbit. The last is inaccessible. To see it, it must become manifest; to become manifest, it has to be transformed. The true subject of discourse is therefore not the path of the electron, but the transformation of this path.

Scientists have thus uncovered seemingly insurmountable barriers blocking the way toward complete knowledge of our material universe.

However, these difficulties do not mean that all attempts at understanding must be abandoned. On the contrary, they stimulate the search for other paths that promise to lead to new and hitherto unexplored domains whose very existence was unknown: science uses its crises to broaden its vistas, by means of new concepts.

A similar process can be detected in the development of the researcher's universally used tool, logical argument, which has also met with unforeseen obstacles.

UNDECIDABILITY

The scientist's permanent tool is logical argument, that is, the application of a certain number of rules ensuring that the conclusion reached is rigorously deduced from the initial hypotheses or facts. One specific branch of mathematics was developed with a view to determining the rules of this logic, and to ensuring that a group of propositions are coherent and based on sufficiently rigorous proof.

Thus, various relationships between propositions within a discourse were defined, relationships that make it possible to affirm the truth or falsehood of these propositions. For instance, the well-known rule of the syllogism: if the propositions P_1 "All men are mortal" and P_2 "Socrates is a man" are true, then the proposition P_3 "Socrates is mortal" is equally true; in other words, P_1 and P_2 imply P_3.

At the beginning of this century, the inadequate formulation of this logic locked set theory into insurmountable paradoxes, linked to an implicit belief in the existence of the "set of all sets." By showing that the "set of sets that are not elements of themselves" obviously cannot exist (since it neither can nor cannot be an element of itself), Bertrand Russell provoked a crisis that led to a quest for more precise axioms (see box 2, p. 105).

An attempt was then made to define a body of axioms (that is, of propositions that would be true whatever the elements involved) allowing one to decide, for every correctly formulated proposition, whether it is true or false. This quest seemed to be a legitimate, even an essential, activity. Nonetheless, in 1931, the Austrian mathematician Kurt Gödel showed that this objective is unattainable. His "theorem of incompleteness" shows that if a body of axioms is sufficiently rich to build arithmetic, the coherence of the system based on these axioms cannot be demonstrated without introducing other axioms. In other words, having adopted such a body of axioms, it will always be possible to find a proposition P that can be proved neither true nor false: this proposition will be "undecidable."

A particularly striking example of an undecidable proposition was found in 1963 by a student of Gödel, Paul Cohen; it concerns a hypothesis formulated in 1878 by Cantor, "the continuum hypothesis." The latter had proved that the cardinal of the set of whole numbers, infinite and all though it be, is smaller than that of real numbers:[2] there are "more" real numbers than whole ones (while there are "as many" even numbers as whole ones, since they can be put in one to one correspondence with each another). Cantor named these two levels of infinity *aleph* 0 and *aleph* 1, and suggested, without being able to prove it, that there is no intermediate level; this is the "continuum hypothesis." Cohen showed that it is impossible to prove, by means of the founding axioms of arithmetic, whether this hypothesis is true or false. Cantor's proposition is undecidable. To use it in an argument, one is obliged to assume it as an additional axiom or, if one prefers—but this is purely a matter of choice—to posit, on the contrary, as an axiom that it is false.

This result is all the more remarkable in that many researchers had, since Cantor's time, tried to prove his hypothesis. It is quite possible that some of the many "conjectures" (that is, unproven statements) proposed by mathematicians are in fact undecidable propositions and that attempts made by researchers to find a proof are totally futile. This may be true of Fermat's famous "great theorem," for which, according to a note the latter wrote, in 1673, on the margin of a book, he had found a truly remarkable proof, but did not have enough space to write it down. According to this theorem, a, b, c, and n being odd numbers, the equation $a^n + b^n = c^n$ has no solution if n is greater than 2; for three and a half centuries no general proof of this has been found. Similarly "Goldbach's problem" still remains one; this German mathematician stated in a letter addressed to Euler in 1742 that every even number is the sum of two primary numbers; this statement, which is anything but intuitive (because the larger the number, the lower the density of the primary numbers, while that of even numbers remains constant), could never be shown to be wrong, no matter how large the numbers studied, but it still has not been proved.

If the undecidability of a proposition of this order were one day to be demonstrated, we would have to accustom ourselves to the idea that our good old arithmetic, so solid, so rich in certainties, also has its own shadowy areas and secret gardens.

Pure logic, in the most formal domain possible—the weighing of propositions—was thus capable of marking out its own limits. It can be seen as a kind of defeat of the mind, which is thus shown to be definitively incapable of forging an instrument capable of catering to all its needs. It can, on the contrary, be seen as a victory: our mind was itself capable of finding the inevitable imperfections of what it constructs, even before having completed it. One can thus feel relieved at finding that the yoke of logic will never close in on us completely: we will always have the freedom, on encountering an undecidable proposition, to accept or reject it, as we choose.

Science, so often presented as an increasingly complex and powerful machine allowing man to improve his understanding of the world around him in order to subjugate it more effectively, has thus secreted, at the very heart of the organs which direct its functioning—the observation of reality and logical argument—these two unexpected and peculiar concepts: uncertainty and undecidability.

This does not mean that this machine has begun to backslide or has suffered just some easily reparable breakdown; it is the true nature of scientific activity that is illuminated. To be sure, this activity leads to a better understanding of the world. It sometimes opens up new possibilities for action, but, as Paul Claudel says at the beginning of his *Art poétique,* it is less a matter of knowledge than of personal growth, that is, of a kind of development that satisfies one of our intellectual needs: taking possession of what surrounds us through understanding it.[3]

2. *Science and Everyday Life*

This review of the internal difficulties of scientific activity is likely to be perceived as an academic exercise, without any repercussions on the actual functioning of the mechanism, something like a casual display of some matter of conscience that has scarcely any relevance to everyday behavior.

For it is our everyday life that has been profoundly affected. Over the past two centuries, scientific discoveries have, at a particularly fast rate, been converted into technical inventions that have changed not only man's environment, but especially how he occupies this environment, how he envisages his own destiny.

For every person and for every group

—the question is to be,
—the question is also to be happy.

This need has become a demand according as the new possibilities accompanying technical advances have given the impression, or the illusion, that it could be satisfied. However, this progress has simultaneously changed the contours of the problem by transforming, in all domains, the very nature of human experience.

MAN AND WORK

The "industrial revolution" created a collective mentality whereby work is linked with profit, profit with material power, material power with happiness. Work, once thought of as the result of a divine malediction, has thus become, through profit, the source of happiness; it is claimed as a right.

However, the very dynamic of technological progress, which increases productivity, makes this work less and less necessary. Instead of finding (or rediscovering) a kind of balance within which working days are in the minority, the immediate reflex is to safeguard employment by creating a tertiary sector which is not only useless but is sometimes even destructive—a bureaucracy which justifies itself by means of the tasks that it allots to itself.

Work has become the main activity of human life; the child is prepared, conditioned for work; the age-group capable of working is privileged; the end of work is experienced as a penalty or a deprivation.

Efficiency and productivity have become the chief criteria for activity. They have fostered a prodigious increase in our capacity to produce riches, but they have necessitated increasing specialization.

Thanks to the accuracy with which information can be transferred between individuals, it is possible to assign a very limited portion of the global task to each one, thus allowing him to acquire an expertise which increases as his field of activity narrows. In a society where efficiency and productivity have become the key words, the process of specialization and of increasing professionalization has accelerated.

The mass of knowledge required for work of any complexity has grown to the point where no one individual can acquire it. Collective efficiency requires that an individual be trained in one narrowly defined area, and that this training then be put into practice through careful orchestration of the different skills.

This increasingly necessary collaboration may, to be sure, stimulate a spirit of mutual help and tolerance, increase awareness of the need for contributions from others, and repress the temptation to despise them. However, it also has some harmful consequences. "Professionalization" leads in reality to

—A compartmentalized society: the specialization of language makes most types of exchange impossible; mediation becomes necessary. *Communication,* that is, sharing, a collective and balanced act in which each individual participates on an equal basis, is replaced by *information,* that is, imposing *form* or shape, an act requiring techniques of various kinds as well as centralizing and broadcasting networks.

—A stratified society: conviviality, which is based on equality not of individuals (which is meaningless), but of the status granted to individuals, cannot develop; relationships are based on the implicit acceptance of the domination of one party and the submission of the other; a hierarchical structure seems inevitable and is accepted as natural.

—Isolated individuals: according as he acquires an ever deeper knowledge of his discipline or profession, the individual becomes confined to an increasingly narrow group whose expertise, language, and obsessions he shares. Gradually, he has contact only with his peers, whose number decreases according as the degree of specialization increases, and communication, even within the family, is affected;

—Mutilated individuals: the development of certain characteristics is accompanied by the atrophy of numerous other ones. Like a weightlifter, the professional is at a loss when confronted with something that is outside his field; he becomes monomorphic, unipotential, unable to cope on his own with everyday situations; above all, the knowledge that his competence is so narrow in its scope makes him anxious about change;

—Frustrated individuals: a few centuries ago, society was no doubt structured into specialized groups (farmers, artisans, warriors), but the domain assigned to each one was broad enough for him to have a sense of personally accomplishing "something"; each person could feel a certain pride in the result, whether tangible or not, of his work. However, this initial specialization gradually changed in character, and we now have a fragmentation such that most of those who participate in the productive process feel that they themselves are "making" nothing.

Finally, for most of us, the aim of work is not what it allows us to create, but the profit that we can make from it. Work is thus considered as a right, not because of the satisfactions it provides, but because it is the means of making profit and therefore of access to wealth. (To quote Marcel Pagnol's Topaze: "Money does not make happiness, but it enables us to buy it from those who do make it."[4]).

MAN AND HIS BODY

There is one domain where this specialization has reached grotesque proportions—that of the body. In our societies, attention is paid only to those functions of the organism that contribute to its efficiency in the productive process or in certain conventional activities like sports.

Most of the body's possibilities are forgotten, atrophied. Even "audio-visual" learning techniques appeal to only two senses, hearing and vision, while the organism is actually much more diverse and rich. Thus, the learning of music, which should be aimed primarily at unfolding a new world of sensation and of interpersonal communication is deflected toward the arid study of the rudiments of music or toward a quest for virtuosity; here too the aim is to be efficient in the race for success. The growing awareness of the body has been channeled toward the training of sports champions, who are put through the same paces indefinitely, under unchanging conditions. This quest for virtuosos and champions requires an elitism that excludes all but a few exceptional individuals. Here again the aim has been distorted: it is no longer a matter of increasing the possibilities of one's own body the better to enjoy it, but of emerging a winner from pitiless competition.

Is it going too far to see much of our behavior as a kind of prostitution? Who is more a prostitute—the engineer who leases his intelligence to improve the techniques and financial returns of an enterprise whose goals are entirely unknown to him or the woman who rents her charms by the half hour for the benefit of a customer of whom she knows nothing?

To characterize this degradation of the role of the body by means of an extreme (but completely factual) example, we will visualize 42nd Street around Times Square in New York. Countless sex-shops and porno-shows advertise "novel" pleasures in attractive and glittering display windows. To gain admittance, one has of course to pay a few dollars which the many poor rubberneckers reduced to window shopping obviously do not have. Fortunately for them, a few branches of the local blood transfusion center are set up on these very sidewalks to buy blood from anyone who is willing to give some. A few cubic centimeters of one's blood in exchange for the right to watch a striptease, that is the deal.

However, in a less ruthless and more insidious and subtle way, how many deals do we not make where the terms are basically very similar?

MAN AND HIS FELLOW HUMAN BEINGS

The most significant technological upheaval is no doubt that which increased the speed of communication. The network for the transfer and storage of information has become amazingly efficient. This new power fascinated at first, but it is beginning to cause anxiety:

—The daily knowledge of all world events, perceived at first to be a description of the reality of the universe, quickly appears to be a camouflaging of this reality. The news items that we hear each day are like flashes illuminating alternately and randomly some small portion of an immense mosaic; the meaning of the whole increasingly escapes us according as these flashes become more rapid and numerous. Real changes in society are usually gradual and result from slow shifts in thinking; they never become "news" since they are not distinguished by any single "event." We are surprised to notice one day that in one particular domain and then another, things have changed without anyone having made a decision about it. But it is too late; we make the best of it. In the last analysis, the development of society is nothing more than a series of unplanned changes to which people resign themselves.

—Saturation prevents us from classifying and ordering news; it creates a refusal reflex and a desire for escape into fiction. A vague and undifferentiated skepticism sets in little by little. Today's citizens, satiated with news, are just as credulous and defenseless as those who were entirely deprived of it.

MAN AND HIS PLANET

Thanks to the technology which we have evolved, we alter almost at will, and will alter still more easily in the near future, the natural conditions under which we live. A short time ago, we became aware of the long-term consequences of this intervention. We discovered that even apart from any nuclear cataclysm that we may trigger, we are exhausting the available resources, slowly accumulated over geological eras, and are upsetting age-old balances.

Our planet is not so huge, and we are condemned to staying on it. Every few seconds, we destroy, often for futile gratifications, riches slowly created by nature: to keep the "public" informed about some champion's moment of weakness or some film star's divorce, we destroy acres of forest to make newsprint, only a tiny fraction of which will be looked at and which, a few seconds later, will have become ashes and carbon dioxide gas. Our accumulated waste products are piling up, and the very oceans are becoming waste-bins, which no garbage collector will ever empty. Our search for instant comfort and pleasure is threatening our long-term survival; it is, at the same time, threatening all the species that cohabit with us on the earth.

One of the most aberrant and spectacular cases of the absurd destruction of a collective treasure is the massacre of whales; all species of this mammal are becoming rarer and are in danger of becoming extinct. The total number of blue whales, the absolute giant not only of the sea but of all species now living or that ever lived on earth, is now only about 3,000, whereas it was more than 40,000 a half-century ago. Long overdue protective measures have been taken, but the two countries most responsible, Japan and the Soviet Union, opposed their application for a long time. The aim of this massacre is particularly derisory: whale flesh is used essentially for the manufacture of dog and cat food, of lubricants and lipstick.

For life to be maintained on this planet, should we, like Nietzsche, wish for and organize the death of man?

TRAINING FOR THE ROLE OF AN ADULT HUMAN BEING

The young of humans require by far the longest learning process. The mastery of all the forms of behavior necessary to living life as an adult human has always taken many years. The accumulation of new knowledge has considerably lengthened this process and, above all, has led in our societies to the specialization of the role of educator. This role is now centered increasingly in the school, unlike so-called primitive societies where it was shared by the entire group.

In a society geared toward the specialization of each of its members and subject to the all-powerful criterion of efficiency, school has evolved according to a natural process which leads to strange absurdities. The goal is no longer to further the development of each individual—it is to supply society with individuals who will fit satisfactorily into the mechanism of production. Thus its primary function, education, has become secondary to its selective and orientational functions. Each child is launched into a ruthlessly competitive obstacle race where blunders are punished with assignment to a more restricted course. The most fortunate do of course get an opportunity for a broad education, but they quickly find that each obstacle successfully surmounted is only a prelude to the next. Not only school but life as a whole has been transformed into a succession of waiting periods; success is never experienced as happiness, but as preparation for the next inevitable exploit. No stage is an accomplishment, and nonetheless the ineluctable final stage is well known.

MAN AND HIS DEATH

Many societies have managed to integrate the scandalous mystery of death into their way of life. With the advance of science and its offshoot, technology, our society's ability to resolve this contradiction is diminishing steadily. Which is better—an absurd death in a car accident, an anonymous death in a hospital, in the cold isolation of medical technology, or a slow death in a specialized home catering for people at various stages of decrepitude?

In supplying a few very partial explanations, in allowing people to believe that everything is explicable, in destroying the ancient myths, science has created a void; it cannot, by its very nature, either found a system of values or refer to any transcendence. For a doctor, death has a precise definition; for a jurist, it means a simple change in status. But for the person implicated, it is a matter of ceasing to be.

3. *Science and Human Becoming*

This assessment may seem harsh, may seem above all to be aimed at the wrong target: why should the failure and mistakes of our society be imputed to science which, after all, enables us not only to take possession of the world, but also to define in advance the limits to our understanding of this world?

It is because science is not just one secretion among many of a human group; it is the very process by which we humans are differentiated and distinguished from other primates.

An animal endures; from time to time, by chance, he stumbles on effective behavior; among higher animals, he is capable of making this behavior a permanent part of the group's heritage; however, for want of an abstract understanding of the reasons for its effectiveness, he does not know how to use his invention as a springboard toward greater success. Ethologists have made famous a female macaque, Imo, living on the island of Koshima, in the east of Japan; according to the researchers present, she noticed, one fine day in 1954, that potatoes washed in sea water tasted better than unwashed ones. Gradually, her companions copied her; potato washing is now customary to the group. However, presenting this as a "discovery" is merely playing with words: it is actually a random cultural mutation, similar to genetic mutations, which, brought about through the blind play of accidents occurring during duplication of the DNA molecule, sometimes contribute new functions beneficial to the species.

Man, for his part, transforms deliberately. Capable of imagining a mechanism at work beyond apparent changes, a process where his senses disclose only a chronicle, he is able to impose his will on the world around him and change it for his own benefit.

The knowledge and especially the understanding which allow him to reduce the behavior of the real world to the interplay of a few parameters of his own invention are at the heart of the mechanism at work in the scientific approach. Imagine that, in the apparent void that surrounds us, there exists something which we label "air" or "gas"; characterize this thing by means of parameters as unimmediate as volume, pressure, and absolute temperature; notice that, under certain conditions, the behavior of this "gas" is such that the product of the first two parameters divided by the third remains constant; use this observation to make machines for our use—all that is science, and the various phases are indissociable. For the machine is the outcome of the theory, but in many cases, the direction in which the theory developed and the slant of the questions which furthered its development were determined solely by society's need for a particular machine.

It is only for linguistic convenience that we isolate scientific activity from the rest of the group's activities; science is constitutive of society just as society is the initiator and producer of science. It is therefore not altogether unfair to impute to one the imperfections of the other.

However, we experience a brutal and painful letdown when we move from a description of the victories of human intelligence, whether over the material world or our own spirit, to a recognition of the impotence, failure, and destruction plaguing a society such as ours, guided by this intelligence. Must we concede that scientific activity, indissociable from our human status, represents a danger for our future? I do not presume to offer a solution to a problem of this magnitude, but to contribute, by means of examples relevant to biology, and more precisely genetics, to certain analyses that may clarify this question.

In the first part, I will deal with some of the pitfalls along the path of our thinking on any subject, whether scientific or not: pitfalls concealed not only in words and numbers, but also in the need to classify the "objects" that we discern or imagine around us. One of the roles of science is to help us to avoid them; unfortunately, the inappropriate use of a superficially scientific approach often contributes to our being prey to them.

The aim of the second part will be to illustrate the true role of science, which is not to answer questions asked, but to imagine pertinent questions.

Finally, I will outline certain necessary (but, of course, insufficient) changes in direction intended to slow down or even, if that be possible, reverse the present rush toward the abyss, of which scientific progress is one of the driving forces.

PITFALLS

Little by little our mind was structured and our intelligence formed and enriched, through contact with a few other people—parents, brothers and sisters, teachers, friends. They fashioned us, sometimes without knowing it, and created intellectual reflexes in us which conditioned the gradual flowering of our personality.

During our adolescence came a time of questioning; we rejected certain ideas, refuted certain passively accepted opinions. After a personal analysis, we adopted new attitudes; we consciously freed ourselves from a part of our origins.

But this questioning did not, in general, extend to the very tools with which we build our thinking. Now the shape of these tools influences, more than one might imagine no doubt, the final result. Elementary concepts such as those of number or of class were impressed upon our minds along with a whole range of properties and ways of handling data. The tools which we use for description, mainly words, have been internalized, as have the grammatical rules governing their associations, to the point of structuring not only our discourse about things, but our very vision of them. Incorporated very early into our intellectual reserves, these tools were scarcely ever subjected to critical revision. We use them out of habit, without paying much heed to the slant which they put on our thought. They are indispensable to us, to be sure, but if we lose sight of the conditions for their use, they may become formidable traps for our thinking. It is to the outlining of such a critical revision that the next three chapters are devoted.

2.
Number Pitfalls

The universe around us becomes present to us through an uninterrupted flow of sensations. Gradually, our mind applied itself to classifying these sensations, replacing the infinite diversity of objects that we can perceive with a more limited collection of groups, or classes, to which we assign these objects.

This activity, which is based to a large extent (we will stress this in the chapter on "classification pitfalls") on arbitrary choices, leads quite naturally to characterizing, in particular, each class according to the number of objects assigned to it. In our apprehension of the real world, number plays such a decisive role that it behooves us to ask whether it is not excessive. It is with number that we began our apprenticeship to "science," that is, to a discourse that strives to give a vigorous description of the universe. Thanks to numbers, we are equipped with a marvelously efficient tool, on the one hand, for ordering and, on the other, for applying rules which enable us to arrive, without ambiguity, at a number from other numbers. But this efficacity harbors pitfalls of which we must become aware.

THE PITFALL OF HIERARCHIZATION

The collection of so-called "natural" numbers follows an order which corresponds to the question "larger or smaller?" and this order is used as a reference in all other orders which we can imagine. Whatever the nature of the objects under consideration, order can be made of them only if one can determine a way whereby the set of numbers can be applied to them (that is, if to each object we assign one number *and one only*). But such an application is possible only if the sum of all the information available about each of these objects can be summarized by means of a single parameter. When I am satisfied that this parameter does adequately characterize the objects being considered, I can, with pertinence, ask the question "Is object A greater, equal, or less than object B?" The answer will be a function of the numbers X_A and X_B associated with the two objects because, in the case of numbers, the question "greater or equal?" is meaningful.

If the objects under consideration are people or groups of people, it is therefore possible to describe them as being superior or inferior provided we specify the way in which we assign a number to each person or group. Naturally, there are countless possible ways of assigning numbers. One can, for example, using well-defined techniques, measure, for each individual, his weight W, his height H, his annual income I, and his intelligence quotient IQ, and assign to him a number X, calculated from these four parameters, and written mathematically as: $X = f(W, H, I, IQ)$; A will be greater than B, if $X_A > X_B$.

Clearly, this way of putting it is extremely dangerous because it is too likely to be interpreted as signifying that one of the individuals is superior to the other when this relationship exists only between the numbers that we have arbitrarily associated with these individuals.

When our knowledge of objects grows to the point where we realize that we can no longer, without betraying them, characterize them by means of a single parameter, we lose the ability to place them in hierarchical order. Immediately after we consider two supposedly irreducible parameters which cannot be condensed into one through choice of one function, the question of superiority loses all meaning. For example, if we characterize each individual by his income and his IQ—A by I_A and IQ_A, B by I_B and IQ_B—the only thing we can do to compare them is to try to establish whether or not they are "equal": thus, we can write $A = B$ only if we find that, simultanously, $I_A = I_B$ and $IQ_A = IQ_B$. If one of these equalities is lacking, then A is "different" from B; $A \neq B$, but there is no question of superiority.

When we are comparing one number to another, nonequality implies that one is greater than the other; when we are comparing sets, it implies only that they are different.

This is not a plea motivated by moralistic considerations; it is a statement of a logical fact. To fail to take account of it is to make a misinterpretation, an error against which, unfortunately, our education arms us but poorly.

The most blatant example of this type of misinterpretation is linked to the finding that people are different from each other. Genetically, that is obvious; the number of possible combinations of the various pairs of genes that we carry is so great that the probability of meeting two genetically identical individuals is nonexistent (except in the case of monozygotic twins); added to this genetic difference are all the variations arising from each person's experiences. Regardless of the criteria that we choose, no two people therefore are ever "equal." This obvious finding leads most minds, often even reputedly brilliant ones, to deduce that certain people are "superior" and others "inferior."

The collection of nonsensical statements made as a consequence of this implicit error—nonequality implies hierarchy—is particularly rich on the subject of human races. Of the many available, let us quote this text by Francisque Sarcey published in 1882, commenting on a work by A. Bertillon, *Les Races sauvages:*

> "These awful bipeds, with ape-like faces, prancing about and voracious, spluttering inarticulately, are our brothers, or . . . the brothers of those who were our prehistoric ancestors! . . .

> Some races, more highly gifted than others . . . set themselves apart from this barbarous animality, became cultivated and refined, . . . formed civilized man, . . . further from a poor Australian than that Australian from a gorilla. Others did not develop; . . . still as lacking in moral sense and in reason. . . . They are the last witnesses to vanished epochs . . . . All these savage tribes are going to become extinct, . . . either exterminated by superior peoples or dying out all by themselves . . . . It will not be a pity."

We may laugh; but the consequences of such absurdities in our century are well known, absurdities pronounced (and this is important) in the name of science. An eminent man like Francisque Sarcey, reputed to be enlightened, debonnair, and liberal, would not perhaps have written these outrageous things if he had been taught early in life that nonequality does not imply superiority.

The trap of hierarchization, set by number, can be all the more subtle in that it is based on the implicit but widely held belief that numbers introduce rigor, that numbers guarantee the scientific character of an argument. A very revealing example is provided by certain passages of *La Race française,* a published work by Dr. René Martial, a member of the medical faculty at the University of Paris, in 1934.

Anxious to give a solid biological basis to the concept of race, Martial states that the most meaningful comparisons must deal not with obvious traits—skin color, skull shape, and the like—but with the genetic structure of populations: two human groups differ fundamentally more with regard to the frequencies of the various genes found within them than with regard to their outward appearance. This approach has since been widely adopted, and it is quite remarkable that a doctor, a nonspecialist in genetics, should at that time have chosen this path which has proved fruitful.

Unfortunately, the data then available for characterizing populations dealt with scarcely anything other than the ABO blood group system. Martial notes that from one group to another the frequencies of the various phenotypes,

A, B, AB, and O, are very variable. Thus, in Europe, the B group is more frequent in the east and especially in the south-east. To characterize the structure of a group, he uses a ratio baptized the "biochemical index of blood, I" defined by the equation

$$I = \frac{f_A + f_{AB}}{f_B + f_{AB}}$$

where f_A is the frequency of the A phenotype.

Why not? This index is somewhat strange, but it allows us, it is true, to differentiate populations rich in group A, with a high "index," from populations rich in group B, with a low "index." The author lines up numerous numbers, but what is remarkable is the gradual transformation of index I, initially a simple marker, into a criterion of value: the index of French people is 3.2, that of Germans is 3.1 (but it is as high as 4.9 in the Black Forest region, where the "inhabitants are brown brachyocephalics of the Alpine type"), that of Poles is "only" 1.2, that of "Jews" only 1.6, that of "Negroes" only 0.9. A long argument attempts to lessen the bad impression created by the Poles' low index: students in Warsaw schools reach 1.6, so "the Polish index is therefore much closer to ours than one might have thought. Franco-Polish marriages yield very good products. However, it must be recognized that . . . some of the grandchildren return to Poland, just as Indochinese mulatoes return without fail to yellow. There is therefore a possibility of waste."

Let us not dwell on the unfathomable stupidity of these considerations. Our purpose is to show how the use of numbers made it possible to give a dogmatic statement ("The French are superior to the Jews, the Jews to the Poles, and the Poles to the Negroes") the appearance of scientific proof. It was sufficient to think up an index, calculated in so complicated a manner as to disconcert inexperienced minds, and to use this index as a value-scale.

It is true that the "biochemical index" of the French is higher than that of Poles. What is arbitrary is to deduce from this that the French are superior to the Poles. This statement implies that the index measures a value, a hypothesis which obviously has no basis.

All this would not matter very much if authors such as Martial confined themselves to making verbal judgments about individuals, blood groups, or populations. But their aim is to act in order, of course, to "improve" the species.

Having assumed implicitly that every good citizen ought to strive to increase the "biochemical index" of his country, and doing a little more mathematics, Martial notes that to increase a quotient it is enough to decrease the denominator; he therefore proposes to "eliminate" B individuals from the French community and "to only keep the AB if they have a favorable bill of psychological and physical health." Ready for experiments of all kinds, he is also interested in French Canadians, "a vigorous and sturdy race, in spite of a somewhat weak personality and an overdeveloped critical mind," and he proposes to "transplant a certain number of families" (that is, to crossbreed them) to a French (Normandy, Brittany) or a "neo-French (Algeria, Tunisia)" environment.

These lucubrations would not be worth a moment's attention except that they are exemplary. They prove first of all that the racist delirium was not the monopoly of the Germans, and among them of the Nazis. They show above all that by sprinkling one's discourse with a few mathematical terms, one can quite easily succeed in convincing those of one's readers who do not have the training and the time required to scrape off this false facade.

In the case of Dr. Martial, one may assume that he himself was misled by this camouflage. It is difficult to believe in the good faith of those involved in certain more recent cases similar to his, which must now be dealt with.

One summer, a political tendency calling itself the "new right" received considerable media coverage. One of the objectives of this tendency is to combat "egalitarian fanaticism." Starting from the obvious diversity of individuals and groups, the mentors of this new right proclaim that "inequality is a fact" (A. de Benoist, *Le Figaro,* November 19, 1977) and that it is necessary to distinguish "the s____ from the diamond" within humanity (Louis Pauwels, *Figaro-Magazine,* March 15, 1980).

It is no longer a matter, as in Francisque Sarcey's time, of contrasting "civilized races" with "savage races," but of comparing, at the very heart of our society, those who succeed with those who find themselves excluded.

The aim is to justify the transmission of the social hierarchy from generation to generation by claiming that this hierarchy is natural. This claim was given grotesque illustration by a Californian businessman, M. Graham, who at the beginning of 1979 created a sperm-bank exclusively for Nobel prize winning scientists. All the geneticists consulted about this project were unanimous in denouncing its ridiculous and utterly unscientific nature. The new right presented it as an admirable attempt at improvement of the human species (*Figaro-Magazine,* March 8, 1980).

However, the extensive coverage given to this kind of ideology has contributed to the spread of racism (because what is racism if not the acceptance of the idea that within our species certain groups are, as Louis Pauwels so elegantly put it, "s____," while others are "diamond"?), and we now see instances where it declares itself openly and triggers numerous horrifying incidents. These same authors are then quick to protest that they never preached a natural hierarchy between people. It is difficult to believe in their good faith; they were not, as Sarcey and Martial may have been, victims of the trap set by number; they themselves used a fake science based on number to set a trap of their own.

THE PITFALL OF ORDER

The "greater-lesser" concept defined with reference to numbers allows us to order the latter: it creates an "order." This word, unfortunately, evokes for us far more than a simple arrangement where everything has its place. It also suggests organization, harmony ("There, all is *order* and beauty, bliss, calm, and luxury"[1]). Within our spontaneously black and white vision of reality, order, as opposed to disorder, is of course on the right side. Among our Swiss friends, satisfaction is total when it can be said that "everything is in order." Even physicists use this word loosely in reference to the degradation of energy or, to use the appropriate scholarly terms, the growth of entropy. What they are referring to is the process, summed up by the second principle of thermodynamics, by which matter tends inevitably toward ever greater "disorder." The reality summarized in this principle includes the ineluctable ageing of structures, the gradual loosening of the links between the elements which ensured the coordinated functioning of the whole, the appearance of disorganized zones which spread little by little and dissolve the global organization. We will return in chapter 8 to the problems raised by this principle of waning, of whose limits certain authors are currently reminding us.[2] At this point, we wish to stress how ambiguous it is to present it as a disastrous but inevitable triumph of disorder over order: why should a set of particles scattered at random, more or less uniformly, in space be deemed to be less "in order" than a collection of these same particles in a few highly interactive aggregates? The answer is a question of definition or of feeling.

The use of the word "order," given the vaguely arithmetical or unmistakably aesthetic aura attached to it, is often a betrayal of trust (how could one refuse to defend order?).

This pitfall was admirably denounced by Paul Valéry, who wrote: "Two dangers threaten the world, disorder and order." Life strives to steer a middle course between generalized disorder, which signals the onset of death, and absolute order, which signifies the victory of death; between disorder, which deprives activity of goals and words of meaning, and order, which imposes immobility and silence. The two abysses are equally dangerous; one must, like Jean le Bon at the battle of Poitiers, protect oneself as much on the right as on the left.

However, our reflexes make us more attentive to the risks of disorder. We find it normal that the courts condemn those who "threaten public order." If we follow Paul Valéry, we would sometimes have to deal just as severely with those who threaten public disorder. In our satiated and timorous society, the forces favorable to order are quite obviously in the majority; it is therefore necessary to call to mind the symmetrical need for disorder, and when order is embodied in certain dictators of Latin America or elsewhere, we have to remind the authorities that it is their duty to tolerate or even to facilitate a type of disorder which may contribute to the creation of a less menacing order.

THE PITFALL OF ADDITION

The first operation which we learn to carry out on numbers is addition; the others follow. The abstract nature of this operation is quickly forgotten because of the exploits it enables us to achieve when confronted with the problems posed by the real world. Even before knowing how to define a number, we knew that "two and two make four," and we solved problems which were sometimes put in a complicated way but where all the characteristics of the objects were represented by additive quantities, as in the famous problems involving reservoirs and taps.

Later on, our apprenticeship to physics concentrated especially on pressures. Even if we did not understand exactly what was involved, we knew that if two pressures are exerted simultanously on the same point in an object, they "add up"; in other words, it is as if a single pressure, equal to the vectorial sum of the two first ones, were being exerted. As a consequence of this marvelous property, one can, without changing the outlines of the problem, replace one pressure by two or more pressures, provided their vectorial sum is equal to the first; the adept choice of the direction and intensity of these partial forces, the "components," quite often makes it possible to quickly solve a seemingly complex problem—hence the "elegant solutions" for which our high school teachers used to congratulate us.

The real world that awaits us after leaving college does not have the marvelously simple structure of elementary school arithmetic or of the pressures studied in high school physics. We discover areas where the explanation of phenomena does not in the least lend itself to "additive models," that is, models which require that observable reality be broken down into measurable quantities capable of being added to each other. However, the reflex of scientists is then to constrain this reality, to force it to fit into an additive model. This attitude is not absurd; it is even very often effective, at least locally. It allows us to predict and to take action. It nevertheless constitutes a betrayal of the reality which one is claiming to describe and explain. This is particularly clear in biology, and more precisely, in genetics.

The clearest case is that involving the search for an individual effect of each of the genes governing a quantitative trait. In sexual species such as ours, one of the properties of genes is that they are in pairs. We have known since Mendel's time that every elementary trait is dependent not on a single hereditary factor, but on two factors that coexist, unchangeable. They act simultaneously on the trait expressed by the individual; this trait is a result of their interaction and not of the addition of their effects.

Numerous attempts have been made to bring this interaction down to an addition. To illustrate how unrealistic these attempts are, it is enough to imagine a trait with only two possible values, 1 and 0, that is governed by a pair of genes, one of which, *A*, is dominant and the other, *a*, recessive: in other words, in individuals with the *AA* or *Aa* genotype, the trait has a value of 1; in *aa* individuals, a value of 0. Is it possible to attribute to each type of gene a proper effect on this trait? Provided one makes certain arbitrary but legitimate choices, one can indeed reply to this question. But the answer is not supplied by numbers; it depends on the frequencies p and q of the genes A and a; one finds that these proper effects are equal to $+q^2$ for A and to $-pq$ for a. In a population where A and a have the same frequency 1/2, the effect of A is therefore to increase the trait by 1/4, and that of a to reduce it by the same amount, but with the frequencies 9/10 and 1/10, these proper effects are +1/100 for A and −9/100 for a. While the values associated with the two phenotypes are absolute numbers independent of the actual populations, the values attributed to the genes vary as a function of their frequency. These proper effects are not characteristics of the genes themselves, but of the population.

It is in fact impossible, by very reason of the process of interaction, to attribute a specific additive effect to a gene. To be sure, more or less complex methods have been developed to calculate these proper effects. In reality, they can only obtain results that are valid locally in a given milieu, but this limit to their validity is quickly forgotten. The answer obtained does not address the initial question about the effects of A and a, but an entirely different question concerning these effects in a given population. For the sake of the intellectual comfort provided by the additive model, we had to transform our question—it is important to be aware of this.

Again, in the field of genetics, the use, without sufficient precautions, of an additive model led one of the founders of the Neo-Darwinian theory of evolution, Ronald Fisher, to formulate an inaccurate theory.

This theory (which has its strong points but also its limitations) developed initially through attempting to predict the effects of natural selection on the genes situated at one *locus*, that is, governing a single trait. Depending on the genes at this locus, that is, his "genotype," each individual is endowed with a greater or lesser capacity for winning in the "struggle for survival" and for procreating; this capacity is measured by his selective value. Provided one is prepared to tolerate a certain level of arbitrariness, this notion of selective value can be extended to the genes themselves, individually, and to the population as a whole. Numerous results have been obtained in this way, in particular the famous "fundamental theorem of natural selection," demonstrated by Fisher in 1930. This theorem states that the effect of selective pressure is to increase the average selective value of the population. In other words, natural selection, which is so ruthless toward individuals, is beneficial to the population as a whole.

In fact, this formulation is based on confusion between the "selective value of a population," that is, a population's capacity for competing against other populations for the available resources in a certain environment, and the "average of the selective values of the individuals who make up this population," that is, the average of individual capacities for successful competition within a population. It is clear that the capacity of the group as a whole is linked in a nonadditive manner to individual capacities, if only because the group is compelled to create structures requiring different individual capacities (we will return at greater length to this point in chapter 6). There is therefore an implicit switch, due to defective terminology, from one concept to another.

Furthermore, Fisher's demonstration is based on the hypothesis that the selective value of individuals is linked to a single locus. Now, selection acts of course not on a locus but on an individual, that is, on a set of loci. The result expressed by the "fundamental theorem" is valid only if the effects on the selective value of the various loci concerned are all additive. As soon as there are interactions, as is clearly the case in reality, the evolution of genic frequencies may very well cause a reduction, not an increase, in the average. Our total vision of the global effect of selective pressures is transformed by this distinction.

But the area where unconscious reliance on an additive model caused the most misinterpretations and nonsense is that involving questions about the respective roles of genes and of environment in the expression of a trait. Let us attempt to define the terms and to set out the details of the arguments on both sides of this question often designated by the expression "the innate and the acquired."

THE INNATE AND THE ACQUIRED

Every individual obviously results from the action of the "environment" (including all the factors which affected his development: nourishment, education, affection, etc.) on an organism constructed from genetic information assembled, once and for all, during the fertilization of the initial egg. One can, for example, use the term "innate" to denote this genetic information and the term "acquired" for the set of other factors that shaped the individual. If we study a certain trait *T* in this individual, we find that it is the result of multiple causes which can be grouped into two categories—those linked to what is "innate," those linked to what is "acquired." The mathematician then says that *T* is a "function of the innate and the acquired," which he can represent as

$$T = f(I, A).$$

It then seems natural to ask "What in the trait *T* are the proportions of the innate and the acquired?" This seemingly harmless quest for "proportions" may in some cases be meaningful: "What proportions of the average expenditure of a household go to accommodation and to food?" "What proportions of France's energy supply come from oil and from coal?" But usually it is quite obviously absurd: "What proportions of the quality of my television set reception are to be attributed to the transmitter and to the receiver?" "What proportions of the pleasure I get from listening to a concert are to be attributed to the composer and to the performer?"

For the concept *proportion* to be meaningful, it is necessary that the result be analyzable into terms which can be added to each other; it is necessary, in other words, that the "explanatory model" be additive. If it is not, the use of this concept can only be wrong; furthermore, in leading people's thinking astray, it constitutes a veritable trap, which is all the more dangerous for having been innocently and unwittingly set.

The phenomenon "disposable energy" results from the importation of or the production of coal and oil, and these two causes can be added, at least in so far as a common unit of measurement has been agreed upon; one can then claim that oil represents a proportion of x percent. Similarly, since expenditure on accommodation and on food is measured in dollars, one can attribute a proportion of the total to each one. Usually, however, this kind of additivity is impossible, and the question has no meaning. This is obviously the case when the causes under scrutiny are, on the one hand, the genetic heritage and, on the other, the effects of the environment. The function f, to which we alluded earlier cannot, except in very rare cases, be reduced to an addition and written as $T = I + A$, where I and A represent terms depending on genetic factors only or on acquired factors only.

One can attempt to proceed, in spite of all, with the analysis of the phenomenon under study by transforming the question in a way that may seem insignificant but which is, in fact, fundamental. Let us suppose that this phenomenon is characterized by the parameter T; we are no longer going to ask ourselves about the part played by the various causes in determining T, but about their part in determining the variations of T, represented by the symbol ΔT.

Let us be on our guard: we have thus moved from an analysis concerning the trait to an analysis concerning the variations of the trait. The change in the subject of our discourse is considerable. It is not the same thing to speak on the role of the liver in digestion and to speak on the differences in the functioning of the liver from one individual to another. While the first discourse would deal primarily with those functions which are common to all individuals, the second would scarcely mention them.

Strictly speaking, this second question is not, in its generality, any more meaningful than the first, except in exceptional cases. However, if the variations of the "causes," ΔI and ΔA, are sufficiently small for it to be possible to discount those terms that are a degree greater than 2, and if the function T is sufficiently regular, one can accept the approximation

$$\Delta T = a\Delta I + b\Delta A \tag{1}$$

where a and b express the respective "weights" of the variations of I and A in the variation of T.

(A simple example of a "trait" depending on two "parameters" is the height h of a place which is a function of its longitude L and its latitude l. The ordnance survey map describes by means of contour lines the function $h = f(L,l)$. This function is obviously complex; however, if one considers a zone sufficiently small for the ground surface to be assimilated to a single plane, one can write

$$\Delta h = a\Delta L + b\Delta l$$

where a and b characterize the slope of this plane. In solving certain local problems, the use of this formula is highly effective. However, as soon as one moves to broader questions, it loses all bearing on reality.

Thanks to this approach, we bring the variations of T down to an additive model regardless of the complexity of the function f. This is a remarkable success, but let us not forget the price that had to be paid for it: we assumed that the variations of I and A among the individuals being compared are sufficiently small for their squares or their product to be negligible. In other words, we confine ourselves to analyzing the variations of the parameter T within a rather genetically homogeneous group (small ΔI) subject to very few environmental differences. The use of this result to compare groups with very different genetic heritages (large ΔI) or subject to the influence of dissimilar environments—(physical, sociological, educational)—(large ΔA) is therefore fraudulent, in outright contradiction to the hypotheses.

The power of presenting us with a universe where "causes" add up is not the only attraction of equation (1). It offers the advantage of justifying certain observations or certain experiments and of making their exploitation possible. It provides a formulation that is both easy to understand and operational; its success is therefore assured. One may seem a very awkward customer if one recalls its major shortcoming: it has no meaning other than locally.

The attractiveness of a model as simple as that expressed by formula (1) is understandable; its simplicity generates a sense of intellectual comfort; we return to the good old problems of our childhood about reservoirs and taps. The trait under study can be compared to the level of the cistern subject to contributions from two sources, the genetic source and the environmental source; when this level changes by x centimeters, by how much does each source contribute to the change? It must be recognized that for certain precise and necessarily very limited problems this technique can be very useful. However, the use to which it is most often put is so aberrant that it is difficult to remain serene in the face of such an accumulation of errors of logic.

In fact, it is as though having adopted an explanatory model in accordance with the formula

$$\Delta T = a\Delta I + b\Delta A$$

and having then estimated, using more or less complex but usually correct techniques, the parameters a and b, one were to deduce from this that these parameters are constants characteristic of the determinism of T. Now a and b are themselves functions of I and A: they correspond to a local particularity of the function f, but in no sense to a characteristic proper to the unknown mechanisms that link T to I and A.

Thus, the famous and ever so frequently published statement "Variations in intellectual quotient can be explained 80 percent by variations in the genetic heritage and 20 percent by those of the environment" amounts to giving a universal value to the coefficients 80 and 20 when, at best, they can only express the particular case of a certain group of individuals within a certain environment.

Finally, the difficulty with discussions concerning "the innate and the acquired" does not spring from the terms "innate" and "acquired," which, at a cost of a certain arbitrariness, can be given definitions acceptable to everybody; it springs from the term "and," which in many people's minds implies addition, while the reality is a result of interaction.

A well-known mathematical technique, variance analysis, unfortunately makes it possible to answer the question about the proportions of the innate and the acquired without the users of this technique being always fully aware of the limits to the significance of the results obtained. The somewhat esoteric way in which variance analysis is sometimes presented hides the fact that the answer obtained can only be meaningful if the question itself was.

VARIANCE ANALYSIS AND ITS PITFALLS

The function linking the parameter to its various "causes" is generally not known; it is necessary to rely on indirect methods permitting the "explanation" of the variations ΔT by the variations of the various factors on which it depends. Such is the role of variance analysis, which is well known to all those accustomed to statistical calculations. Box 1 explains what a variance is and what its "analysis" consists of.

In order to define the procedure, I will present a very simple case; but I strongly urge the reader to take a pencil and participate by doing the calculations himself (the only requirement is the ability to square two-digit numbers).

Let us consider therefore a trait the measurement of which depends uniquely on the genotype of the individuals and on the environment in which they live. To simplify, we assume that only two genes, a_1 and a_2, are present in the population at equal frequencies, which result in the presence of three genotypes, a_1a_1, a_1a_2, a_2a_2, at the frequencies 1/4, 1/2, 1/4; finally, one half of the population lives in environment I—say, the plain—and the other half in environment II, the mountain. Let us assume therefore that the measure of parameter T, which we assume characterizes a certain aspect of intellectual activity (it is, by the way, the famous "intelligence quotient," but we do not have to say so here, which provisionally avoids all discussion about the meaning of this parameter), is represented by table 1.

Box 1

The Distribution of a Trait: Variance and Variance Analysis

Suppose that we ask each of the students in an amphitheater containing N students to indicate his height: let x_i be the height of student i.

The standard procedure for calculating the *mean* m of these heights is to use the equation

$$m = \frac{x_1 + \cdots + x_i + \cdots x_N}{N},$$

which can be written in more compact form,

$$m = \sum_{i=1}^{N} x_i/N,$$

the summation sigma meaning that one makes the sum of the x numbers for all the indices of i from $i = 1$ to $i = N$.

Two amphitheaters with the same mean m can of course have very different distributions. One of the characteristics of this distribution is its dispersion around the mean; in amphitheater A, the height of most students is close to the mean and the dispersion is small; in another B, half the students are rather big and half rather small, so the dispersion is large.

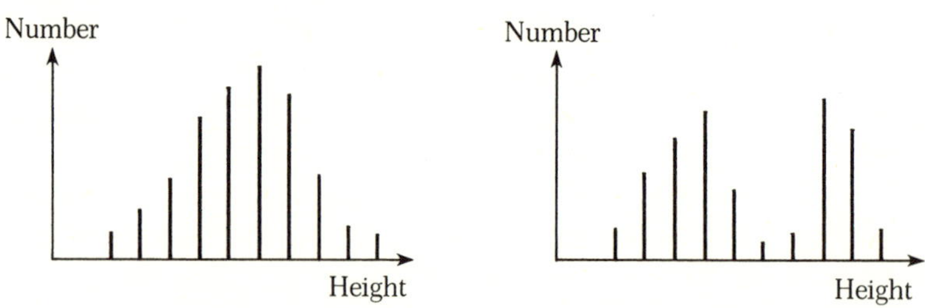

One can imagine different characteristics allowing one to measure this dispersion. The most standard is the variance V, defined as the mean of the squares of the deviations from the mean:

$$V = \sum_{i=1}^{N} (x_i - m)^2/N.$$

If all the x_i are close to m, the variance is small; if some are very far from m, it is large.

An amphitheater containing either boys only or girls only will in general have a smaller variance than an amphitheater with both sexes, because girls have a mean m_G which is lower than that of boys m_B; the presence of people of different sexes therefore causes a certain additional dispersion.

One is thus led to analyze the total variance in two parts:

—one which is the mean of the variances found, on the one hand, among the boys and, on the other, among the girls—it is the variance *within* classes;
—another which is the variance of the means of the two groups—it is the variance *between* classes.

In general, it can be shown that whatever the criterion used to distribute the individuals into classes (sex, color, etc.) the total variance is the sum

—of the variance of the means of the different classes,
—and of the mean of the variances of the different classes, which can be written

$$V = M(V_c) + V(M_c).$$

If the means of the various classes are all equal, their variance is zero, and $V(M_c) = 0$; the dispersion of the trait is entirely due to the dispersion within each class.

If the variances within each class are all zero, $M(V_c) = 0$; the global dispersion is entirely due to the difference between the various classes.

Between these two extreme situations, one can characterize the degree to which the existence of various classes contributes to the dispersion of the trait by the ratio $V(M_c)/V$.

Table 1

Environment	Genotype		
	a_1a_1	a_1a_2	a_2a_2
I	100	140	60
II	70	90	110

In environment I, the simultanous presence of the two different genes increases T, while in environment II, the gene a_2 causes a steady increase in this parameter. This highly realistic example is close to what is actually observed in the case of the genes responsible for certain abnormalities in the hemoglobin, depending on whether the environment is or is not malarial.

The mean for the trait in the population is equal to 100;[3] the dispersion around this mean is measured by the variance; here, one finds that $V = 750$.[4] This variance is due to two causes:

—all individuals do not have the same genotype,
—all individuals do not live in the same environment.

To evaluate the "fraction" of the total variance that is due to genetic effects, it is sufficient to group together all the individuals living in the same environment and to calculate the variance V_G of T for this group; since the effects of the environment are the same for all individuals in the group, V_G represents the dispersion due to the activity of the genes; one obtains here:

in environment I: $V_{G1} = 1{,}100^3$,
in environment II: $V_{G2} = 200$,

or, on average, since one-half of the total population lives in environment I and the other in environment II, $V_G = (1{,}000 - 200)/2 = 650$.

In other words, according to this calculation, the degree to which the genes contributed to the total variance is equal to $650/750 = 87$ percent; by contrast, the degree to which the environment contributed can be calculated to be 13 percent.

This argument seems rigorous and allows us to state "scientifically" that the variations of the trait in question are essentially governed by "innate" factors, which may have serious consequences for many decisions.

Unfortunately, we can approach this analysis in a different but equally rigorous manner and reach the opposite conclusion. To evaluate the proportion of the variance that is due to genetic effects, it is sufficient to group together all individuals with the same genotype (from both environment I and environment II) and to calculate for each one the mean of parameter T; one obtains:

G	a_1a_1	a_1a_2	a_2a_2
m	85	115	85

The dispersion of these means is caused by the action of the various genotypes; their variance, which can be easily calculated, represents the part of the total variance that is explicable by "innate" causes; one obtains $V'_G = 225$; the part played by the innate in the variation of T is therefore $225/750 = 30$ percent; by contrast, we evaluate the part played by "environmental" causes at 70 percent.

The same set of facts therefore leads us, by two equally reasonable paths, each corresponding to a natural intuition, to results that are in total contradiction.

What happened? Where is the pitfall?

It is at the very basis of variance analysis; this involves making one parameter constant so as to measure the residual effect of the other, which can be achieved

—either by calculating the mean of the variances within a constant environment,
—or by calculating the variance of the means corresponding to the various genotypes.

An extremely simple mathematical argument shows that these two paths are equivalent only if the effects of the two causes, in this case the genotype and the environment, are additive—that is, if the effect of each is independent of the modalities of the other. This would, for instance, have been the case if the table had been as shown in Table 2.

Table 2

Environment	Genotype a_1a_1	a_1a_2	a_2a_2	Effect proper to the environment
I	95	125	95	+10
II	75	105	75	−10
Effect proper to the genotype	−15	+15	−15	100

In a case like this, environment I increases T by +10, regardless of genotype; the a_1a_1 genotype reduces T by −15, regardless of environment. There is no interaction. Readers can check that this time the two paths which we described lead to identical results. But as soon as this additivity can no longer be assumed, any analysis in terms of proportions is totally useless since it leads to conflicting results, all of which are equally defensible. In effect, in our initial table, the variance of the effects of interaction was particularly high since the various genotypes led to very different traits depending on the environments. The total variance (750) could be analyzed into

—the variance of the proper effects of the genotypes $V_G = 225$,
—the variance of the proper effects of the environments $V_E = 100$,
—the variance of the interaction effects $V_I = 425$.

It is a pity that these obvious mathematical facts, which any first-year student of statistics could grasp (all that is required is the ability to calculate means and variances), should have escaped so many psychologists and that the famous little sentence specifying the percentage to which IQ variations are due to environment and to genes, which we have already quoted, continues to be pronounced. It is not a matter of questioning the numbers being put forward but of becoming aware of the hypothesis that must underlie them to give them meaning: genes and environment should simply add their effects to each other. Few specialists would dare to claim that such a mechanism exists. No matter how frequently this little sentence is repeated (and it was again recently in a very successful book by a French child psychiatrist), it can therefore only be rubbish.

In a still famous apostrophe, André Siegfried, wishing to make people aware of the actual geographical structure of our planet, threw out this advice: "Let us demercator ourselves!" Our image of the world was indeed deformed by those maps where Mercator's projection gives undue importance to the Canadian north or to Siberia and strangely reduces the land surface of Africa; it is therefore necessary to forget Mercator.

Is it not even more important for us to "de-additionalize" ourselves?

There is no question of denying the usefulness of this operation; the proposal made in jest one day to a few top officials from the Department of Education, to introduce it only in the final grades and even then with great care, is, to be sure, extreme.

However, it is by no means futile for us to become aware of the dangers of a tool which we manipulate with excessive ease and to which we are perhaps introduced too early. Let us conclude with an anecdote which is strictly true. One morning somewhere in the Senegalese bush, my companion was completely happy: he was going to be able to marry the girl he loved. "Why did you wait till now?—I had to give her father a cow.—How did you get it?—With seven goats; here, a cow is worth seven goats.—How did you acquire the seventh goat?—With six chickens; here, a goat is worth six chickens." In order to impress him, I replied: "Therefore, to you, a cow is worth forty-two chickens," but highly amused at my foolishness, he said, "No one would be so silly as to do that."

Indeed, what a ridiculous idea to wait until one owns forty-two chickens, impossible to count and to transport, before buying a cow!

I had learned very young that one should not add up cabbages and carrots; I had just learned that one should also exercise caution before adding chickens and chickens.

3. The Pitfalls of Classification

The reality which we perceive around us is a mass of disparate objects, all unique and exceptional. Grouping them together as a function of common traits sometimes necessitates strenuous exertion of our mental faculties. Each of the luminous spots which we can see in the nocturnal sky has a certain individuality; we are able to recognize the same ones night after night. It is, however, quite natural to see them as a category of objects, stars. It requires a vigorous stretch of the imagination to include in this category another luminous object, but one which is quite different in appearance, the sun; it requires still another stretch to eliminate seemingly identical objects—planets—from this category.

For our knowledge of the universe to progress, it is necessary, given the limitations of our minds, to replace the infinite diversity of reality with a far more limited number of categories or classes. In order to classify, we are obliged to retain only a fraction of the characteristics which we are capable of distinguishing in things—we have to impoverish our vision. But at that cost we are capable of creating a certain order, of highlighting certain relationships between objects.

This order, however, is not intrinsic to the things themselves, but is a facet of the vision, or rather the representation, which we have of them. The "reality" which we are addressing and within which we are trying to identify patterns and interactions in order to gradually develop a "scientific" discourse is but a caricature, arbitrarily created by our mind from the "reality" grasped by our senses.

To prevent this caricature from being a betrayal and to keep it as faithful as possible to the nature of the objects under consideration, some precautions are necessary. Our spontaneous mental reflexes in no way guarantee that these precautions will be taken; we can hardly feel more reassured concerning classifications which we have not made ourselves, but which we have accepted throughout our upbringing, particularly while learning to speak, because all language implies classification.

We were gradually moulded by rigorous discipline, necessary to our communication with others, designating things first by spoken, then by written words; a very small proportion of these words are "proper" nouns, applicable to a single object, and the vast majority are "common" nouns applicable to categories which include innumerable undifferentiated objects.

"Scientific" endeavor has been pursued in the same direction; does science not strive to define efficacious categories, to gradually replace proper nouns with common ones? For the radiant orb of day, whose whims dominate our daily life, is substituted a very ordinary star, the seat, like many other similar ones, of totally banal elementary nuclear reactions.

We have become accustomed to this mechanism which allows us, certainly, to develop our understanding of the universe and even to assure our power over it, but which, by the same token, uniformizes reality by replacing the originality of each object by what is common to a category. This increases our sense of intellectual comfort, but the price exacted is high. What is most unfortunate is that we become insensitive to the dangers of this process and may slip into applying methods of category definition in areas for which they are not valid. It is therefore essential to clearly define this mechanism, to specify the hypotheses on which it is based, and to draw the limits of its significance. Of what then does the intellectual process which allows us to classify consist? On what arbitrary concepts is it based? Even if our answers seem obvious, an awareness of the details of this activity is useful.

Let us note first of all that the word "classification" designates both the actual process by which a classification is made and the end result of this process. Moreover, two phases must be distinguished in this classification:

—a first one which consists of defining the various classes: taxonomy,
—a second which consists of assigning each object to a class: identification.

These activities have been particularly well studied in so far as they pertain to living beings; we will confine ourselves here to this particular case.

THE ARBITRARY CHOICE OF TRAITS TO BE CONSIDERED

A thing considered in its own right, in its totality or entirety, cannot be classified; it is there in front of me, irreducible, oblivious of my categories; more essence than form, it escapes me. To grasp it, to fit it into the well-ordered classes which I have defined or am about to define, I must forget it and replace it with a group of characteristics that I have previously chosen. I must forget Valy, my faithful companion, and content myself with observing that his coat is such and such a color, that he is such and such a weight, has such and such a type of snout, such and such a way of barking or running; based on this, I can decide that he is an animal, a dog, a German shepherd, and can thus assign him to a category.

Depending on my objective, I will choose, in order to make this assignment, sometimes one group of characteristics, sometimes another. I cannot therefore classify objects, but only groups of traits measured from these objects.

For a long time the characteristics taken into consideration were ones which could be directly apprehended by our senses—color, shape, weight, behavior; the data used therefore concerned only *phenotypes*. Thus, within the mass of living creatures, one can isolate those with an internal skeleton, the branch made up of "vertebrates"; among these vertebrates, those that nurse their young, the class of "mammals"; among mammals, those whose encephalon is most highly developed, the order of "primates"; finally, among these primates, all those individuals with whom we are interfertile, the species "man." At each stage of the classification, the criteria used are linked to one or more easily observable traits.

It is well known, moreover, that in their enthusiasm taxonomists did not stop at the level of species, which often include a considerable number of individuals, but analyzed the latter into subspecies, into "races" and "subraces" (the term "sub" obviously does not correspond here to a value judgment, but simply means that the categorization is extended to the next level). Here again, phenotypic criteria were used—skin color, height, hair texture—which allowed the definition of three standard "human races": yellow, white, and black.

But advances in biology have shown that traits which are not directly visible can have still greater importance in the description, and therefore the classification, of a living being. Landsteiner's discovery in 1901 of the first blood group system, the ABO system, demonstrated that all humans belong to four categories (since subdivided, but here we are going no further than this level of precision): A, B, AB, and O. These perfectly clear-cut categories allow unambiguous classification. Gradually, many other systems were discovered, especially the very rich and diverse ones concerning the immunoglobulins (the Gym system) or histocompatibility (HL-A system).

Knowledge of these characteristics has reached such precision that, thanks to them, every individual can be rigorously identified, so great is the number of possible combinations between the many modalities of the various systems.

Moreover, these characteristics are independent of the life experience of the individual from conception onward; directly governed by the genes received from his parents, innate, they remain immutable throughout his lifetime. The data used no longer concern the phenotype, more or less at the mercy of environmental influences, but the *genotype*, which is strictly stable.

This stability, this direct link with the biological heritage received by the individual and as a function of which he has developed, makes genotypic traits most reliable for the establishment of a taxonomy and for the assignment of the objects under study to various classes. Unfortunately, the available data usually pertain to phenotypic traits, which are the only directly observable ones. It is possible to derive one from the other only in those rare cases where an unambiguous phenotype-genotype correspondence could be established. Even in the case of the ABO system, which is especially simple, this correspondence

genotype	*AA*	*AO*	*BB*	*BO*	*AB*	*OO*
phenotype	*A*	*A*	*B*	*B*	*AB*	*O*

does not allow one to deduce the genotype from the phenotype; an A individual is just as likely to be *AO* as *AA*.

The difficulty is even greater in the case of quantitative traits for which one is capable only of constructing mathematical models, for want of knowing the biological mechanisms by which the genes influence the apparent traits: thus, in humans, skin color is transmitted "as though" the intensity of the pigmentation was dependent on four or five pairs of genes functioning in an additive manner; in reality, the determinisms involved are probably much more complex and linked to a much greater number of genes, but they are still unknown.

These models, which supplement our ignorance, have illustrated the operational efficacity of certain concepts such as "heritability," making it possible, in a global way, to link phenotype and genotype. The traits used as a basis for taxonomy and identification are therefore no longer directly measured, but are estimated by means of probability distributions.

This recourse to probabilities is necessary, moreover, when the objects under consideration are not individuals, but groups of individuals—populations. These populations are known to us only through more or less representative samples, allowing us only to estimate the probabilities of their various traits.

Finally, the starting point for a classification is a list of objects (for example, the mass of all known living individuals or the ensemble of populations which they constitute), opposite which we place a list of traits, these being either measures or parameters subject to standard laws of probability.

It is clear that, depending on the state of knowledge, depending on our investigational skills, depending also on our a priori opinion as regards the importance of the various criteria, the list of these traits can vary widely.

ARBITRARY CHOICE OF A "DISTANCE" BETWEEN OBJECTS

When "objects" are characterized by a single criterion, it is easy to group together those that are similar: classifying according to height or to weight presents no problem. All this changes, however, as soon as one wishes to take account simultaneously of two criteria or more—to classify, for example, both according to weight *and* according to height. Now, in order not to betray the objects under study too badly, we obviously must take the greatest possible number of criteria into account.

Comparing two objects i and j, whether individuals or populations, involves comparing two ordered groups of numbers,

$$x_i \{x_{1i}, x_{2i}, \ldots, x_{ni}\} \text{ and } x_j \{x_{1j}, s_{2j}, \ldots, x_{nj}\},$$

where x_{1i} is the measure of trait 1 for object i. We then find that our minds are incapable of replying simply to the fundamental question "Does object i resemble object j more closely than object k? "a question which can also be phrased "Is i closer to j than to k?"

The use of the term "close" leads us to speak of "distance"; any classification involves, in the last resort, defining a measurement system, imagining a space where the objects under study are represented by points; similar objects will be represented by points which are close together. For the mathematician this space is quite simply, the hyperspace defined by coordinate axes that are equal in number to the traits which we considered in these objects. It is then necessary to decide on a system of measurement, that is, to adopt a method of calculation that makes it possible to obtain a number d_{ij}, the distance between i and j, as a function of the elements of the vectors X_i and X_j.

Mathematicians, with their very fertile imaginations, have invented numerous techniques, all of which are perfectly justified but which sometimes lead to very different results, for calculating these d_{ij}'s.

The most famous is "classical euclidian distance," whose square is equal to the sum of the squares of the differences between the measures for i and for j; this is the distance which we used in our youth when we applied Pythagoras' famous theorem.

Very useful sometimes is "Manhattan distance" where d_{ij} is the sum of the absolute values of these squares (which does indeed correspond to the distance to be covered between two points in New York: the difference between the avenues plus the difference between the streets).

More sophisticated, "Mahalanobis distance" takes account of the links between the various characteristics (information on height gives information on weight since these two traits are correlated); defined in 1936, it requires the inversion of the variance-covariance matrices between the measurements, which made it impractical until rapid calculators were made available to researchers.

Particularly prized by population geneticists, "arccosine distance" (the distance between *i* and *j* is the angle whose cosine is equal to the sum of the products of the square roots of the frequencies of the various alleles) is very useful in the comparison of populations as a function of the content of their genetic heritages.

Parallel to these "distances," coefficients of similarity or dissimilarity were defined, playing analogous roles (for instance, Karl Pearson's famous coefficient of racial likeness), and with various advantages or disadvantages.

This list has but one goal—to show that the definition of a system of measurement is no trivial matter. The same data may lead in certain extreme cases to totally conflicting "similarities" or "dissimilarities" depending on the particular formula used to calculate the distances between the objects. In actual fact, this danger seems, in nonpathological cases, more theoretical than real: the use of the various techniques usually leads to very similar results.

The choice of a particular distance is usually dictated by the habits of the researcher or by the availability of serviceable calculation programs rather than by a theoretical analysis of their respective merits. It is useful to remember this in order to increase one's awareness of the relative nature of certain discussions.

Most methods for the calculation of distances require, at the outset, a reply to a new question: must the various traits under consideration be weighted, and how? It seems in effect that certain criteria, either because they are measured with a greater degree of precision, or because they have a smaller dispersion, or, above all, because they correspond to characteristics deemed a priori more important, must weigh more heavily than others in the global distance.

Endless discussions have been focused on this problem. It was found to be impossible to objectively define the "importance" of a trait (cf. Sneath and Sokal 1973), with the result that many specialists assume that it is preferable to grant equal weight to the various parameters, whatever they may be.

The problem that we are now addressing is anything but academic; depending on the weightings adopted, Pygmies could be closer to Eskimos than to people from the Nile (because of their height) or the opposite (because of their skin color).

Similarly, in comparing various populations, one can for equally good reasons give particular importance to differences between the frequencies of rare genes, or, on the contrary, to those at average frequencies, or to those that are very common; the results may be noticeably affected, as was shown by S. Karlin's recent study comparing various Jewish populations.[6]

Any classification should therefore specify not only on what criteria it is based, but also what relative importance was attached to each of them, and what technique was used to synthesize the set of differences into a "distance."

ARBITRARY CHOICE OF A METHOD FOR DEFINING CLASSES

The group of objects under study, individuals or populations, having been represented in a space provided with a system of measures, all that remains to be done is to regroup them into more or less homogeneous and distinct subsets. Two methods can be followed: a "descending" one which involves successive separations and an "ascending" one which involves successive agglomerations.

Our mind's spontaneous reaction usually favors the use of a descending approach. This was the case in the example mentioned earlier, beginning with the animal kingdom and ending with the human species. When confronted with a set of numerous objects, we create groups as a function of a criterion, putting to one side all those for which the criterion has modality *X*, to the other all those for which it has modality Y (white people and black people, for example); then we further analyze each of these groups as a function of another criterion, etc., which allows us to draw a classification "tree" that is gradually ramified as in figure 1.

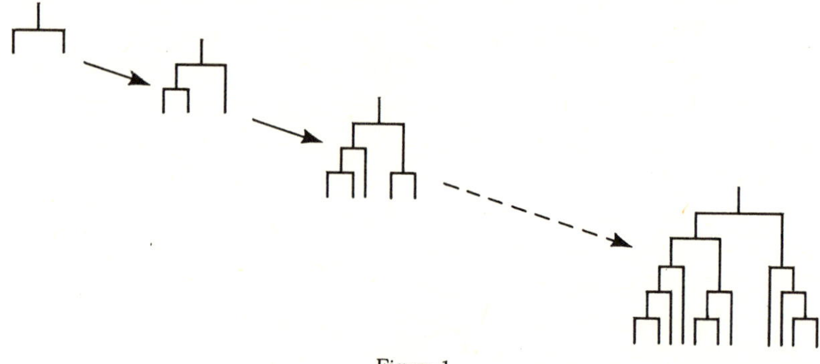

Figure 1

The process stops when all the criteria have been explored. The classes at each stage of this process are called "monothetical," because each one is homogeneous for the group of traits taken into consideration upstream.

The result obviously depends on the order in which the various criteria are considered; different orders can lead to completely heterogeneous classifications. The latter are therefore less a reflection of the natural order of things than a consequence of the hierarchy assumed a priori in the order of these criteria. To reduce this arbitrariness, taxonomists Williams and Lambert proposed to give priority to that trait which is the most closely correlated with all the others and which can be detected by a simple calculation, but this is again an arbitrary approach.

In theory, it is possible to carry out a "descending" analysis without creating a monothetic class by taking account at each stage of the set of traits by means of a global distance: it is sufficient to choose which of the distributions of the n objects to be classified maximizes the sum of the $n_1(n - n_1)$ distances between the n_1 objects which are assigned to one class and the $n - n_1$ which are assigned to the other.

However, to choose this distribution it is necessary to calculate this sum of distances for all possible distributions. Now the number of possible distributions is $2^{n-1} - 1$, or, if $n = 50$, close to a million billion. This operation must be carried out at each stage of the tree's ramification; even the fastest computers would never succeed in classifying, under these conditions, more than twenty objects. We are greatly deluded therefore in thinking ourselves capable of carrying out a classification of this kind on the strength of our intuition alone.

The most frequently used automatic classification methods are therefore not descending, but, contrary to our spontaneous approach, ascending: they proceed by grouping similar or close objects into classes which are, in their turn, compared to the other objects or to the other classes so as to be regrouped. This procedure may be represented by figure 2.

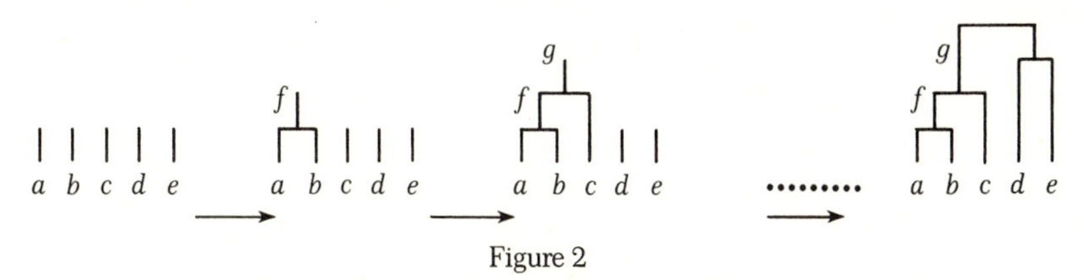

Figure 2

The two closest objects *a* and *b*, were put together in class *f*, then class *f* and object *c* which was closest to *f* were put together in a class *g*, and so on.

Innumerable techniques have been proposed for drawing this gradual outline of a tree whose main branches are defined starting no longer from the trunk but from the farthest boughs. Partial reviews of these techniques are to be found in Sneath and Sokal (1973) and in Belabre (1971). The results can vary greatly depending on the algorithm adopted. In practice, researchers (with scarcely any concern about overusing computers) analyze their data by means of various methods, explore the same method while varying the parameters as a function of which the regroupings are made, sketch the different trees obtained, and select that or those which seem most "reasonable." This may lead, ultimately, to accepting an a priori opinion and to reinforcing it with a glut of calculations.

There is no denying that these calculations very often make it possible to extract a structure from a set of seemingly totally chaotic data, to pinpoint some associations, and to reject others; they constitute a tool allowing us to "see" reality better. Nonetheless, it is necessary to remain aware of the limitations of this tool, and especially to avoid placing greater confidence on the results the more esoteric the underlying mathematics and the more costly the computer used.

PHENETIC TREES AND PHYLOGENETIC TREES

The trees obtained, whatever the techniques used, are "phenetic" trees, that is, they take account only of the traits manifested (directly or otherwise) by the objects under study.

When these objects are living beings or groups of living beings, we know that they have a genealogy, the history of their ascendants. Is it possible to reconstruct this history from their classification, to draw a "phylogenetic" tree?

The initial hypothesis necessary is that the similarity between two individuals or groups is all the greater the more numerous the genealogical links between them, that is, the closer the relationship between them.

When dealing with sexual beings who are by definition descended from two parents, a major obstacle arises immediately: their genealogy is a network and not a tree in the sense in which we have defined it. The expression "an individual's genealogical tree" is completely inappropriate, since it can only be established by reversing the direction of time, by representing the most recent individual as the trunk and the distant ancestors as the branches.

The search for a phylogenetic tree can be meaningful only in the case of groups of individuals whose differentiation occurred through a series of successive splits without any fusion. Thus, it is useful for depicting the gradual differentiation of species throughout evolution, since each species is defined by the existence of barriers causing reproductive isolation (at least among animal species—in the case of vegetables or of microorganisms, a species can be created through the hybridization of two others, which leads to a "reticulated" evolution and not to evolution in the form of a tree).

Studies attempting to reconstruct the history of the differentiation of species have multiplied since advances in biochemistry have made it possible to compare identical proteins (hemoglobin, cytochromes, etc.) in different species, and to compare them at the finest level, that of the sequence of amino acids which constitutes them. A convergence between the paleontological and the biochemical data is gradually emerging, which can be considered a remarkable success for the disciplines in question.

This success may lead certain researchers to infer phylogenetic trees from phenetic trees without always taking the necessary precautions. This is what happens when, based on the classification of humans alive today as a function of their various morphological or genetic traits, people attempt to define "races" and to outline the history of their differentiation.

THE DEFINITION OF HUMAN RACES

The definition of races, initially based on their apparent characteristics, must in fact take into account only those biological factors that are really transmissible from one generation to the next, that is, genes. It is no longer a question, as in the nineteenth century, of differentiating groups of individuals according to their apparent traits, their phenotypes, but according to the contents of their genetic heritages. The following definition of race, which appears in the most recent work on human genetics, has met with unanimous approval:

"A race is a group of individuals with a major part of their genes in common and which can be distinguished from other races by these genes" (Motulsky and Vogel 1979).

There remains the need to assign a content to this definition by specifying which genes distinguish the "groups of individuals."

As it happens, the trait which had originally led to a first classification, skin color, is subject to a strict genetic determinism. In effect, it is less a question of color than of quantity; the dark appearance is due to a pigment, melanin: while it is present in the skin of black people, it is absent or very scarce among white- or yellow-skinned people. This difference in genetic structure can be explained by the effect of natural selection which operates as a function of the intensity of ultraviolet rays. Vitamin D, which is necessary for bone calcification (its absence causes rachitism) is made in the skin under the influence of these rays, which penetrate more easily if melanin is absent. In Europe and in northeast Asia where UV rays are less intense, individuals lacking melanin are at a selective advantage; the genes responsible for production of this pigment have gradually disappeared (this explanation is, however, unsatisfactory in some special cases, such as the Eskimos and Pygmies, who, living in the far north or in the shade of the forest, receive only a small quantity of UV rays but nonetheless have very pigmented skin). A first classification of humans into two groups can therefore be carried out as a function of the genes responsible for melanin synthesis (these genes are still poorly known, but their number can be estimated to be four or five pairs, that is, genes situated in four or five loci, this term designating the spot on a chromosome where the genes governing an elementary trait are located). One can thus divide humanity into "black" populations, on the one hand, and "white" and "yellow" ones, on the other.

Another genetic trait also makes it possible to divide humanity into two groups: the persistence of lactase.

In most mammals, milk contains a carbohydrate, lactose, whose digestion requires the activity of an enzyme, lactase. During lactation this enzyme is highly active, after which it slows down greatly, causing lactose intolerance in adults. In certain human populations, on the contrary, lactase activity persists at a high level (75 percent of that of newborns) during the entire life span, and no lactose intolerance occurs. This trait, linked (it would appear) to a pair of genes, is very widespread among northern European populations and a little less so in the Mediterranean region, but it is rare in Asia and in Africa. (Imagine the consequences of this difference for health improvement programs in certain populations. What is good for Europeans is not necessarily good for Asians or Africans.) This time the classification of humans into two groups as a function of the frequency of the genes involved would single Europeans out from people on the other continents.

Finally, let us consider two biological traits whose genetic mechanism is well understood—the rhesus blood system and the HL-A immunological system.

The rhesus system is governed by genes situated in three loci, each including (if the various rare variants are overlooked) two categories of genes; eight combinations are therefore possible. One of them, called R_o, is present at a high frequency only in black Africa; another, called *r,* is very rare in Asia and in the Pacific, but has a high and noticeably constant frequency from one population to another in Africa and in Europe.

The HL-A system is linked to four loci occupied by very diverse genes. An analysis of all the available data from forty-eight populations made it possible for M. Greenacre and L. Degos to define relatively homogeneous "clusters," one of which includes the European and African populations, another the Asians and Eskimos, a third the populations of Oceania.[7]

Finally, these two systems lead to a classification into two groups, Asians and Eskimos, on the one hand, and Indo-Europeans and black Africans, on the other.

Depending on the criteria chosen—skin color, lactase persistence, or immunological "systems"—our vision of the relationships between the three most commonly evoked large human groups is totally changed: we can arbitrarily say that group A is differentiated from groups B and C which are similar and justify our statement with a biological argument, no matter what groups are designated by A, B, and C.

In other words, as shown by figure 3, we can arrive at three classification trees for these groups depending on the criteria chosen. This result is a consequence of the absence of a history of humanity expressible in the form of a gradually ramified tree. This history consisted of a network with as many exchanges and fusions as splits; it is therefore illusory to try to specify a classification when it cannot have a global meaning.

However, instead of this method of descending classification through successive separations which we have just used, one may choose an ascending one through the grouping of globally similar populations.

Knowing the frequencies of the various genes in the various populations, one can calculate a distance between two populations taking account of the set of differences found between their genetic heritages. The definition of races then consists of looking for groups of populations such that the distance between two populations is small when they belong to the same group and large when they belong to two distinct groups.

It turns out that, for the human species, this method is ineffective.

To demonstrate this, we recall the results obtained by R. Lewontin and M. Nei:[8] they found that 7 or 8 percent of the average genetic diversity of our species is explicable by the differences between nations belonging to the same race and 85 percent by the differences between populations belonging to the same nation. This result can be expressed by saying that the distance between, say, two French populations is, on average, smaller than the distance between two white populations chosen at random, but only by 7 percent; it is smaller also than the distance between any two populations chosen at random on the earth, but only 15 percent.

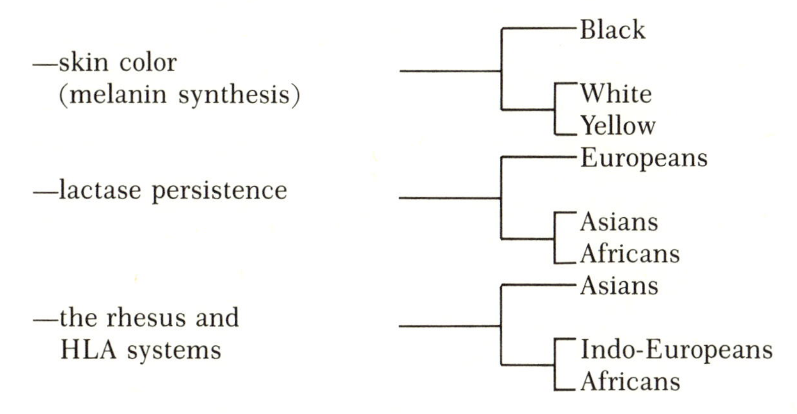

Figure 3

These differences between groups are so slight that the result of any classification is at the mercy of the traits being considered and of the classification techniques adopted. To illustrate this instability, one can, as Cavalli-Sforza and Edwards have done, compare two classification trees obtained, on the one hand, from data on the various blood systems and, on the other, from anthropometric measures.[9]

Numerous inconsistencies appear: the Eskimos, close to the Indians and the Maori in one tree, are close to the French and the Swedes in the other.

There is therefore no question of denying the differences between the various human groups: a black-skinned African is able to synthesize melanin, while a European is unable to do so; an adult European maintains lactase activity, while it disappears in most Asians, etc. But the combination of similarities and dissimilarities is so complex that the picture becomes blurred as soon as one strives toward a vision that would take all the available data into consideration.

To be sure, people are different, but precisely because of sexual reproduction, this difference is much more apparent between individuals within the same family or the same population than between families or populations. My neighbor is genetically different from me; his belonging to another village, another nation, another "race" distances him a little more from me but the additional difference is in each case so slight that it does not provide a basis for drawing truly meaningful frontiers between groups.

A geneticist therefore has a clear reply to questions about the content of the word "race": this concept, in the human species, does not correspond to any objectively definable biological reality.

CLASSIFYING AND MULTIDIMENSIONALITY

The concept of "distance" was introduced to resolve a fundamental difficulty: how can two objects characterized by numerous measures be compared? By means of a distance, this multiplicity of measures is replaced by a single, more or less arbitrarily defined one; one can then regroup the closest objects, in the sense thus adopted for the idea of proximity.

An entirely different and less arbitrary approach can be used, which mathematicians call "multidimensional analysis." Let us attempt to describe the principle underlying it. Objects characterized by a single measure can be represented by points situated on a straight line, after having chosen a starting point and a unit of length on this line; if they are characterized by two measures, this representation leads to points on a plane with two coordinate axes, whereas objects with three measures are represented by points in space. In this mode, it is convenient to consider that objects characterized by 4, 5, . . . , 10 measures are represented by points in "spaces with 4, 5, . . . , 10 dimensions." Unfortunately, we do not know how to see such spaces (and mathematicians are no better than the average person at it—it is not that a highly complex mathematical concept is involved, but purely a facility of language).

To enable our eye to focus to the best of its ability on the position of the points representing the objects in these "hyperspaces," we project the latter onto visible spaces, that is, with 1, 2, or 3 dimensions; this projection is done in such a way as to deform the cloud formed by these points as little as possible.

This is accomplished using methods which are quite simple theoretically, but which require such laborious calculations that they were not used at all prior to the era of electronic calculators; now, all that is required is to have access to one of the innumerable programs that have been developed for this purpose, and to press a button. After which, all that remains to be done is to interpret the results intelligently.

An example drawn from a French political event will show the usefulness and the limits of these methods, but above all it will demonstrate the dangers of an intuitive classification.[10] The event in question was the 1969 presidential elections; there were seven candidates at the first ballot: Messrs. Defferre, Ducatel, Duclos, Krivine, Poher, Pompidou, and Rocard. An examination of the voting trends in the 31 constituencies of Paris allows up to determine how each of these constituencies viewed the candidates: the latter are "objects in a 31-dimensional space," these dimensions being simply the number of votes obtained in the 31 districts. We wish to stress that there is no question of comparing the candidates as persons or their programs, but only the way in which the electors votes placed them relative to each other.

To anaylze the positioning of these seven points in the simplest way, it is sufficient to project them onto a straight line (that is, a one-dimensional space). The result is remarkable (figure 4): at one end we see the point "Duclos," at the other the point "Pompidou," toward the center the point "Poher," between him and the point "Duclos" there is the point "Defferre." This corresponds to the classic left-right opposition with the representative points lining up in order: communist, socialist, centrist, gaullist. But this nice order is disturbed by the three other candidates, Krivine (Ligue révolutionnaire), Rocard (Parti socialiste uni), and Ducatel (independent); the corresponding points are close to center-left.

Since this result hardly corresponds to general opinion, the analyst concludes that projection onto an axis, which reduces the number of dimensions considered from 31 to 1, impoverishes the available information excessively. He then projects the initial space onto a plane; in other words, he reduces the number of dimensions from 31 to 2 by considering a second axis perpendicular to the first.

This time the result coincides with the predictions (figure 5): on this second axis, the first four candidates and Ducatel are located close together, while the points "Rocard" and, even more so, "Krivine" are very far away.[11] In this second direction, it is no longer a question of left and right but of an entirely different characteristic that sets two of the seven candidates apart from the five others. The labeling of this characteristic is a matter of personal choice; depending on

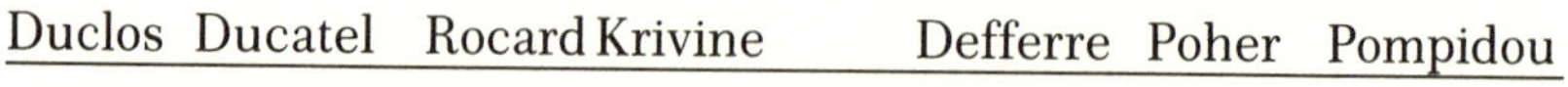

Duclos
Ducatel
Rocard
Krivine
Defferre
Poher
Pompidou

Figure 4 Unidimensional projection

Rating of candidates in the 1969 presidential elections

Duclos
Ducatel
Poher
Pompidou
Defferre
Rocard
Krivine

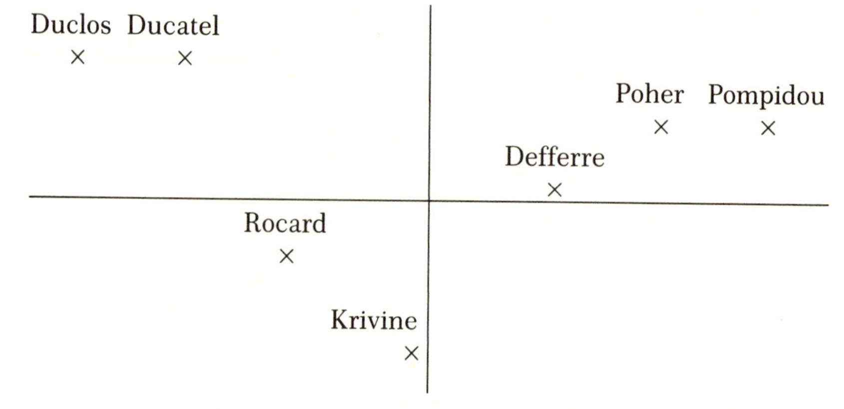

Figure 5 Two-dimensional projection

individual preferences, one can say that it distinguishes the "old-timers" from the "moderns," or those in favor of order from those in favor of disorder. The mathematician has no say in the choice of interpretation. He shows us that the usual distinction between left and right is meaningful only when comparing certain candidates; it is inoperative otherwise.

Spontaneously, our thinking about politics is unidimensional, classifying parties, their leaders, or the people to whom we are speaking according to whether they are more or less to the left or to the right. Fortunately, opinions or doctrines are in reality far less clear-cut; a mathematical analysis shows that other oppositions, other perspectives must be taken into account. In fact, this type of analysis can be even more useful: it allows us to classify the various perspectives in order of importance, and even to say the extent to which each of them contributes to explaining the differences found between the various objects being observed (in our example, the candidates).

We will not dwell here on lines of argument that rely on a somewhat complicated methodology. Let us say, to illustrate this approach, that the analysis of Parisian votes in 1969 showed that 86 percent of the differences perceived by the electors between the candidates were explained by a "left-right" perspective and 7 percent by an "old-timer–modern" perspective.

These global values mean that for the principal candidates (those who received the highest number of votes), the one-dimensional left-right analysis is sufficient, but that for the others, it completely camouflages the reality: Messrs. Ducatel, Rocard, and Krivine were, in the minds of the electors, neither on the left nor on the right; it is as a function of an entirely different criterion that they did, or did not, vote for them.

What is true of political analysis is also true of most domains where we need to classify. In order to get our bearings amid the mind-boggling diversity that surrounds us, to reduce everything to a single perspective or dimension is such an impoverishment that, in essence, reality escapes us; we see but a caricature of it. With two dimensions, we can hope to come closer to an accurate understanding of the group of objects about which we are claiming to form an opinion. But this requires considerable mental effort, and our intellectual laziness often leads us to adopt a simplistic vision where left is set against right, black against white, ugliness against beauty, evil against good.

Adapting André Siegfried's remark about the necessity for "de-mercatorization," we have already noted that to avoid the pitfalls of numbers, we needed to "de-additionalize"; it is no doubt even more important, to avoid the pitfall of simplistic classification, to "de-unidimentionalize" ourselves.

Our intention here was not to denigrate an intellectual process, classification, which prevents us from being submerged by the constant tide of information coming from the outside world. We do not have a choice between classifying and not classifying; because of our own limitations, we have to classify, categorize, recreate the external world, make it ours in a sense, by imagining entities out there, and by naming them.

But in this activity the element of choice is, by contrast, huge, and we must remember that. The way in which I categorize, classify, and name our universe is a function of the entire cultural fabric within which I communicate with others. An Eskimo or a Pygmy confronted with the same necessity will respond in a different way, which results in formidable communication problems between us.

As it happens, this activity has, in our culture, been subjected to a systematic analysis which has made it possible to replace our own intellectual tool with new tools, computers, which perform better in certain limited domains. An entire scientific discipline, "automatic classification," has been developed; its paraphernalia of equations, algorithms, and programs confers it with great prestige and intimidating power. We risk giving the results obtained through these techniques a confidence proportional to their complexity; quite to the contrary, we must remain critical and verify in each case the validity of the underlying hypotheses.

It is crucial to remain conscious of this obvious fact: any classification, no matter how banal, is arbitrary.

4.
The Pitfall of Words

It is with words that we express our thoughts and communicate with others, but the role of words is not limited to that: they participate in the very elaboration of these thoughts. When we try to clarify within ourselves some vague idea that is hovering dimly in our consciousness, we automatically have recourse to words; it is they that give form to this thought, but which also elaborate its content. In the course of the difficult and sometimes painful search for this "idea," an extraordinarily complex process takes place of which we are only partially aware; it stirs up evocations, associations, images, but above all, words. Often, the result of this process surprises us when the sentence that we finally articulate is much clearer and richer than the inner movements which preceded it.

In considering words only as a means of expression, of communication, we reduce their role to that of a passive tool, while they participate actively in the process by which a vague inspiration crystalizes, often laboriously but sometimes like a flash, into a coherent thought.

From this double function of words, as means of expression and as means of elaboration of thought, ensues a double pitfall: the obvious one of confusion in communication, where words can become a source of wrong meaning; the more deep-seated and, for that very reason, more dangerous one of imprecision in thought, an imprecision which threatens it with inner contradiction or, worse still, absence of meaning; words are then the source of countermeanings and nonmeanings.

"Proper nouns," which are attached exclusively to a single object or person, escape these dangers; they are by nature precise. "Common nouns" can create the illusion of accuracy; in fact, no matter how concrete the object to which they refer, they always designate a concept, never a thing. "Pebble," "ocean," "house," and "man" are concepts; language is irredeemably impotent to describe a particular pebble, ocean, house, or man, whose richness defies words.

These are all obvious facts that have been proclaimed thousands of times, but nonetheless our daily experience shows us that these facts are systematically forgotten: how many violent arguments could have been avoided if the words used had the same meaning for all those involved; how many useless sentences would never have been written if their authors had stopped to check the meaning of words! It is a salutary exercise, one which often leads to unexpected discoveries, to analyze a word and to clearly define its use. Any word chosen at random could usefully be subjected to analysis of this kind. As a kind of experiment, and in the hope of encouraging readers to carry out this exercise on a few of the key words of their own field of concern, I am now going to analyze two substantives and one adjective—and, to begin with, a word that we misused just a few lines ago, the word "chance."

1. "Chance"

All knowledge is limited, all information partial; every decision is therefore a gamble. The future is uncertain; when faced with any given situation, it is generally impossible for us to predict what is going to happen with certainty; at the very most, all we can do is to list the possibilities. What actually does happen is one of these possibilities; it was "chosen" from among them. But the verb "choose" is, as we learned at school, a verb of action—it requires a subject. What is this subject? It is easy and customary to use the word "chance" to denote it.

Quite naturally, without realizing it, we therefore give the word "chance" the following definition: "Chance is the subject of the verb to choose in the sentence: 'Reality is chosen from among the possibilities.'" A definition such as this one, which really corresponds to our spontaneous approach, may seem insufficient; it suggests that this word is quite often only a cover up, a means of hiding our inability to precisely define the nature of "that" which chooses. Very basically, it is our vision of the transformation mechanism at work in the universe that is involved; it is serious to avert our gaze, to hide behind a word.

PROCESS—DETERMINISM—CHANCE

When we look at the universe, whether it be at the stars that populate the sky or at the minute organisms that populate a tiny drop of water, we see a world that is constantly changing. The objective of science is, of course, to describe this reality with the greatest possible accuracy and detail—to count, measure, weigh—but it is also, and even more so, to explain its transformations. The accumulation of *data* has as its only ultimate goal to allow us to develop explanatory *models*. Observation supplies us with information on the successive states of a particular star or bacterium, it provides us with a *chronicle* of events. Our imagination allows us to propose a *process* explaining, in the simplest manner possible, the causes of this succession.

The essential ingredient in the conception of these models is the concept of causality. Our experience shows us that certain sequences always occur: if I apply a pressure P to an object with a mass m, its movement is characterized by an acceleration γ such that $P = m\gamma$; we conclude from this that the pressure is the "cause" of the acceleration. Similarly, increased heat is the "cause" of the expansion of a gas, or increased pressure the "cause" of its contraction.

The aim of science has been to define concepts making it possible to describe reality by means of measures (mass, volume, pressure, temperature) such that the cause-effect relationships can be described by formulae which are all the more appealing for being simpler.

This chain of cause and effect participates in a universal determinism that profoundly satisfies our mind. Discovering, in the chaos that surrounds us, causal chains that we can describe in simple terms constitutes a triumph of our imagination all the more striking in that it allows us to predict and, in certain cases, to act.

The hope of scientists, especially in the eighteenth century, was to achieve in the distant but conceivable future a total and perfect knowledge of all the determinisms in the universe, permitting infallible prediction of the instant to come. We now know that this ideal is unattainable.

The only attitude coherent with the imperfection of our information is to evoke the future only by enumerating the possibilities (when they lend themselves to being enumerated) and applying ourselves to assigning a probability to each one, that is, a number characteristic of our confidence that it is one possibility and not another that will be realized. The outcome of recognizing our inevitable "uncertainty" in the face of reality is therefore the use in our explanatory models of probabilistic reasoning. This kind of reasoning is a technique, gradually refined since its invention by Pascal, which allows us to maintain the internal rigor of an argument in spite of the incompleteness of the available data. Thanks to probabilistic reasoning, the information available, even when incomplete, can be used to the best advantage.

It is thus entirely possible to take account of the uncertainty without evoking the concept of chance: that of probability, much more easily defined, being within the domain covered by scientific discourse, is sufficient to assure our coherence.

Reference to an agent called "chance" is not really necessary. It is sufficient to admit that a particular event, which we are unable to explain or predict using deterministic mechanisms, has among its various particularities that of being "aleatory"—which obliges us to describe the partial knowledge that we have of it by means of a "law of probability."

One can find analogies to this avoidance tactic for dealing with difficult concepts. For instance, when we examine the movement of heavenly bodies, it is sufficient for us to assume that, in accordance with the law discovered by Newton, what happens is that the masses attract each other with a force proportional to mm'/d^2. It is not absolutely necessary to link this fact to the existence of a "universal gravitation," which is but a word to denote a unifying concept, reassuring to the mind, but difficult to justify.

DEFINITIONS

Designating by a word "chance" the subject of the verb "to choose" when we do not know the real nature of this subject risks creating an illusion: having pronounced the word, we are tempted to believe that it corresponds to an object or a subject, or even to an agent endowed with something like a will of its own. But naming is not sufficient for knowing; there is still a need for definition. Now the definition of chance seems very ambiguous.

One of the most famous definitions is that of philosopher Augustin Cournot for whom "chance is the meeting of two independent causal series." One can illustrate this conception with an incident transposed from Cournot: hunger impels me to go out and head for a bakery; simultaneously, the rain makes a tile slip from the hands of a tiler; it falls on my head. This event is due to "chance," because the causes of my presence in the street and the causes of the tile's fall are independent. However, this definition brings us back to that of the concept of independence which, in turn, is far from easy to define. In a totally deterministic universe, can absolute independence exist? (Within the vision of a universe produced by an initial big bang and subject to an unremitting determinism, no one particle is independent of another.)

One can look for a definition more clearly underlining the link between the concept of chance and our inability to elucidate the mechanisms at work in a process. One can assume, for example, that "chance is the set of factors that intervene, or seem to intervene, in the modification of a system, but whose mode of action we are unable to express in the form of a functional equation between the values of the parameters characterizing the state of the system at a given second and these values at the next second."

This definition has the advantage of bringing to light the fact that the role attributed to chance depends not only on the level of our comprehension of the mechanisms involved, but on the parameters which we have chosen to describe reality. If I describe the evolution of a perfectly isolated gas taking only its pressure, volume, and temperature into account, I can link the parameters at two given moments with the functional equation

$$\frac{PV}{T}(t_1) = \frac{PV}{T}(t_0),$$

and chance has no part in my comprehension of reality. If, on the contrary, I consider the positions and the speeds of all the molecules contained in this gas, I am unable to specify an equation linking these parameters at two different points in time; I am therefore forced to introduce the concept of chance.

THE "LAWS OF CHANCE"

The scientist can thus easily extricate himself from the web of uncertainty by abandoning his examination of the causes of this uncertainty and by confining himself to perfecting ways of thinking that take account of the aleatory aspect of the processes being studied. By attributing probabilities to the various possibilities, he draws the consequences of the available facts; he therefore completes his task satisfactorily. Unfortunately, it has become customary to use the word "chance" in reference to these aleatory events without specifying the definition being given to it. This lack of precaution is a source of futile arguments, a typical example of which is that provoked by the much used expression "the laws of chance." "How," writes mathematician Joseph Bertrand, "can one presume to speak of the laws of chance? Is chance not the antithesis of all laws?"

In fact, this expression is very flawed—it results from confusing chance with aleatory process. The throwing of a dice is an event subject to chance: let us repeat the throw many times under identical conditions; we find that the result of each throw remains unpredictable but that the average frequencies of the various results calculated from all the throws already carried out oscillate with an amplitude that is constantly decreasing; the frequency of result 3, for example, gradually tends toward 1/6, and this result is observed at each series of throws that we undertake. It is as though the capriciousness of chance, which is obvious throw by throw, were replaced by a predictable order, a law. Gradually, the "law of large numbers" binds chance down so tightly that it takes on all the appearances of determinism.

In reality, it is not chance itself that is subject to laws: at each stage of the process it remains just as free, just as vigorous. The laws apply in reality to the experiment in which we are engaged, and they change depending on the nature of this experiment.

If we draw a ball from an urn containing 100 white balls and 100 black balls, each time replacing the ball drawn, the "law" states that the frequency of whites will tend toward 1/2 and even specifies, as a function of the number of draws, the probability of exceeding a given difference between the frequency actually observed and this limit of 1/2.

But let us proceed to do another experiment: starting with an urn containing one white ball and one black ball, let us draw one ball, replace it with two balls identical to the one drawn, and begin again. This time, we can say nothing about the final composition of the urn: after 1,000 draws the urn may contain either 1, 2, . . . , *x* or 1,001 white balls; all we can say is that these results have the same probability. Is it that chance is more potent in this second experiment that in the first? Obviously not; simply, this second process is such that the "law" of large numbers does not apply,[1] while it does in the first.

The supposed "laws of chance" correspond to a certain dependence between the successive stages of the chronicle that we are observing; this dependence results from the very mechanism of passing from one stage to another and not from the intervention of chance. To be accurate, one should therefore speak of "the laws of aleatory processes" and not of "the laws of chance."

A METAPHYSICAL QUESTION

For the experiments that we have evoked—drawing a ball or throwing a dice—it is clear that a better knowledge of the initial conditions would allow us to predict the result with total accuracy. Knowing the shape of the dice, the location of its center of gravity, the pressure exerted on it, and the resistance of the air, it would be possible to determine without error the side on which it will fall. The chance that we evoke apropos this throw is a *reducible* chance, whose intervention lessens according as our information increases; its role may even disappear past a certain threshold in the precision of the information.

Matters are much less clear at the microscopic level: how can one guarantee that the future of an elementary particle is completely defined by its present state and by the state of the environment that surrounds it? We know that we will never, no matter how much our knowledge advances, be able to achieve a capacity for perfect prediction. But, for all that, can we affirm that the future of an object is not rigorously determined by its present state?

The answer does not seem to be suitable material for reflection or experimentation. It does not depend on some still undiscovered property of the physical world; it is properly metaphysical. Each individual is free to assume that he is part of a Laplacian type of universe, where everything is rigged, where all transformation is constrained by strict laws which lead inexorably toward a predetermined future, where everything was fixed from the first moment. He is free to believe, on the contrary, that he belongs to an undecided whole, where the particles are constantly hesitating, capable, like a tightrope walker, of leaning to the left or to the right or of remaining balanced.

In the first vision, time is but a superfluous parameter, since both the future and the past are contained in the present; time is just one dimension among others. In the second, it is the essential raw material of the gradual unfolding of a world perpetually in the making.

The way we see ourselves individually or collectively is, of course, different depending on our choice; in the second vision—and in it alone, it seems to me—we can find room for freedom and for hope.

But science provides nothing to sway us in one direction rather than the other. Similarly, there are two ways of presenting the prodigious biblical definition of God—"I am Who am"—which I was taught as a child. Absorbed by God, time is abolished; the present engulfs and destroys all duration. But there is, it seems, a more faithful translation of which Henri Atlan recently informed me: "I will be Who will be." Is God himself in the process of becoming?

2. *"Intelligence" and "Giftedness"*

We are all proud of the various qualities that others recognize in us or that we recognize in ourselves; the quality that we prize most is "intelligence." What counts is to be deemed "intelligent" or even "very intelligent." By this word we denote all those facets of celebral activity which, whether because of their nature or their level or intensity, are specific to man. Intelligence is the quality par excellence; it is on it, by tacit but very widely accepted agreement, that we build a hierarchy of individuals; our admiration for a gangster whose actions we abhor or for someone in power whose decisions we condemn remains almost intact when we can add "Yes, but he is very intelligent." What do we mean by this?

Specialists on the functioning of the mind, psychiatrists or psychologists, have for close to a century wanted to make their discipline, so full of nuance and subtlety, a "scientific" discipline, that is, worthy of respect, credit, and money.

To do this, they have introduced numbers.

They have succeeded beyond their dreams, since many of our contemporaries think that intelligence is measurable like temperature or pressure; it can be characterized by one or many parameters, whose value the experts try to determine. To this end, they have devised measurement techniques known as "tests": will these tests tell us all that is to be known about intellectual capacities?

ANSWERS AND QUESTIONS

Before continuing, it is appropriate at this point to make a remark which I consider essential. Regardless of what tests the fertile imaginations of psychologists dream up, they all of necessity consist of a set of questions designed to be answered: thus the speed and accuracy of the reply can be observed. But our intellect does not have as its only function the answering of questions; is its most remarkable activity not, on the contrary, to think up questions? What needs to be verified therefore is not that the questions are "accurate" but that they are pertinent and formulated in such a way that they can eventually be answered. As for the speed with which we developed a question, it is usually of little importance. When it is questions rather than answers that are being evaluated, the two criteria accuracy and speed lose their pertinence.

The inner process which eventually culminates in a question is usually complex, subconscious, influenced—usually unknown to us—by all our contacts with the thinking of others. Slowly a personal framework falls into place; problems of deep concern to us are turned over repeatedly in our minds; gradually they emerge from the realm of vague questions, anguishing in their vagueness and therefore repressed into our subconscious; they lead to clearly formulated questions to which a reply can be given.

INTELLIGENCE AND SPEED

Is the slow work of elaborating a question not infinitely more characteristic of our intellectual activity than the more or less rapid discovery of a reply? A question, much more so than an answer, can be new and original.

A personal misadventure made me aware of the importance of this slow process during which our mind assimilates, ponders, and finally elucidates a problem. One morning, for no apparent reason, a truly sharp and, in my view, particularly original idea flashed across my mind, making me feel "very intelligent." That afternoon at the end of a meeting I did not resist the temptation of proclaiming this new fundamental truth to a few colleagues. Instead of the anticipated compliments, one of them replied with a mocking smile. "You do not find this idea interesting?"—"Yes, of course, but it is stated in full in my thesis." Eighteen months previously I had been a member of his jury; I immediately got my copy of his thesis from my library. In no time we found the passage expressing "my" idea almost word for word; in the margin, I had written "no, wrong."

Perhaps I am especially slow—eighteen months to understand a sentence—but I had, after this long delay, come to a full understanding of the problem posed, so much so that I had made it an idea of my own. Would I have been more or less "intelligent" if I had understood this sentence within a week but had not assimilated it? To comprehend is also to grasp, to absorb; of what relevance is speed to this process?

In our perpetually moving society, it is undoubtedly very often useful to have rapid reflexes. It is not wrong to predict greater success for a person who answers questions more quickly than others. But that is just one aspect among others of our intellectual activity; it is given priority only in certain types of culture or in certain circumstances. The peasant of bygone days had to make decisions requiring a lot of "intelligence," but living at the rhythm of the seasons, he could mature them slowly, and speed was hardly necessary.

All the tests carried out in the presence of an examiner equipped with a chronometer are mainly speed tests: this quality is of course important, but why should it be considered the primary quality?

INTELLIGENCE AND "INTELLIGENCE QUOTIENT"

In most discussions involving psychologists the initial discourse concerns intelligence, but very quickly there is no longer a question of anything other than the results of tests, or even the mere synthesis of these results, designated by the term "intelligence quotient." After a few quick sentences allowing that IQ and intelligence are two different objects, psychologists go on to use these terms as though they were interchangeable.

That IQ may eventually be a useful measure is quite possible, but it becomes terribly suspect when one notices how systematically the ambiguity surrounding its significance is maintained.

The I of IQ refers to intelligence only if one is willing to accept a surprisingly arbitrary and restrictive definition of this word. The reply given by Binet, the inventor of the first tests, to the question "What is intelligence?—It is what my tests measure" has often been presented as a quip. But it is not in the least a quip; it is a logical necessity if one wishes I to be the initial of intelligence. It is useful to know that in every discourse where the term IQ appears, the word "intelligence" has a very restricted meaning, very far removed from that of ordinary discourse.

Beauty, just as much as intelligence, is difficult to define. Based on a certain number of "tests" designed to measure ear and nose lengths, head, chest or hip circumferences, softness of skin on the cheeks, and eye color, one could imagine calculating a "beauty quotient." Each individual could compare his BQ to those of Catherine Deneuve or Alice Sapritch, Michel Simon or Alain Delon. Why not? The result, however, would scarcely be impressive, since few people would equate the B of BQ with the many evocations associated with the word beauty.

As for the letter Q, it really means quotient only in the special case where the mental age of a child is compared to his actual age. A division of this kind obviously has scarcely any meaning for an adult (is someone who, at fifty, thinks like someone forty years older more intelligent than average?). Also, the calculation of a quotient is replaced by an indication of the place occupied by the individual under study within the range of the population to which he belongs; by definition, he obtains an IQ of 100 if he is average, of 115 if he is surpassed by 16 percent of this population, of 85 if he surpasses 16 percent. A quotient is therefore no longer what is involved, but rather a marker situating each individual within a set of scores whose distribution has been made, through various devices for weighting partial scores, to conform to a "normal law."

It is, to say the least, surprising that the term "intelligence quotient" has become so widespread when each of its components corresponds to a misuse of language.

The advance of knowledge consists of abandoning certain concepts and replacing them with new ones; for things to be quite clear, the words used should also be renewed at each step forward. If at the outset some researchers genuinely thought it possible to characterize "intelligence" by means of a quotient, their successors gradually abandoned this claim; they modified both their objective and their techniques; would that they had also transformed the words that they use!

DEFINITIONS

Let us endeavor to distinguish the various concepts involved. That with the simplest definition is the level of success achieved in tests developed by psychologists (and whose usefulness we are not discussing here); it concerns a performance requiring our intellectual tool, carried out under conditions that have been standardized and measured with a degree of precision that is, in fact, quite small, but which can be estimated. This level of success characterizes our capacity for solving problems or for reacting appropriately in those situations which the tests mirror. One can, if one wishes, designate this concept by the term *measurable intelligence*. Its principal attraction is that it is characterized by one or more numbers, and therefore lends itself to arithmetical manipulations: classification, calculation of averages, variances, and correlations. Even here, one needs to check that the numbers obtained do validly lend themselves to these calculations, which is far from always being the case. This "measurable intelligence" is what Binet referred to in the reply that we quoted. It is clear that here we are in the presence of a tautological definition which reduces the definition of an object to the measurement of one of its properties, and which, consequently, completely ignores all its other properties. This "intelligence" bears but little relationship to the evocations spontaneously linked with this substantive.

A vaguer concept is that of available intelligence. It evokes all those capacities which our intellectual tool possesses, enabling us to react in the various domains and situations where it is solicited. This intelligence is multifaceted since it concerns behavior, that is, all those reactions triggered by the challenges or aggressions of our environment, whose diversity is unlimited. Gradually moulded by all our personal experiences, this characteristic of what we have become, provisionally, can be denoted by the term *developed intelligence*. Impossible to measure, it can at most be subjected to a qualitative analysis intended to specify its principal traits.

But this intelligence was gradually built on a biological base, the central nervous system essentially, which, as we well know, was itself constructed from a certain genetic heritage. It is then tempting to evoke this genetic heritage's capacity for achieving, under optimal environmental conditions, a nervous system permitting the highest possible degree of intellectual activity. This potential intelligence would correspond to gifts of nature, gifts which we have brought more or less to fruition depending on the ups and downs of our existence. The interest shown in this third definition of intelligence springs from the fact that it corresponds by its nature to something which is transmittable—but a more precise analysis makes this concept very suspect. Evoking it is the same as assuming that a certain intensity or type of intelligence can validly be associated with the collection of information assembled through the fusion of spermatazoa and egg, as assuming that the "gifts" received by each individual at the moment of conception can be specified.

GIFTS

To be sure, whatever in our organism is active, transforms, or secretes does so as a function of the instructions present in the initial egg; a particular antibody fighting to preserve our intactness and a particular hemoglobin chain adjusting itself to other chains so as to participate in the making of a red blood cell result from the precise expression of some information inscribed in coded form on the chromosomes which we received from our father and mother and of which one copy was supplied to each of our billions of cells. This is true in particular of the cells of our central nervous system and of the environment in which they find themselves. Should these pieces of information, genes, produce organs incapable of functioning properly for want of the proper chemical synthesizing product, all of our cerebral activity would be adversely affected. Several genes have been identified whose presence can, in this way, lead to severe mental retardation or even to total idiocy.

The most widespread seems to be the gene for phenylketonuria which, in Europe, affects roughly one child in 14,000. The presence of two copies of the gene responsible for this disorder leads to the gradual destruction of the brain. In the early 1950s, it became possible, through understanding the mechanism involved, to develop effective treatment. In many other cases, on the contrary, it remains impossible, no doubt provisionally, to combat genetic fatality. To be endowed with such genes is truly to have received a negative "gift."

Contrariwise, genes to which exceptional intellectual aptitudes could be attributed have not been identified. Whether in the field of mathematics, of music, or of painting, these aptitudes are an obvious fact, but one whose origin can by no means be specified. In presenting them as "gifts," in describing the children who have them as "gifted" or even as "supergifted," we assume implicitly that these capacities are linked to a genetic heritage, a hypothesis for which up to now no trace of proof has been found (we will return to this subject in chapter 5).

To be sure, genes can make us blind or deaf, and blind children are unlikely to become great painters or deaf children great musicians. Aptitude for drawing or for music is therefore at the mercy of the genetic heritage. But there is no symmetry between superaptitude, on the one hand, and incapacity, on the other. Similarly, there is no symmetry between the causes of extreme proficiency in running and the causes of impotence. The use of the word "gift" is not neutral; it suggests that the exceptional intellectual qualities that we admire in certain people are a direct consequence of their genetic endowment; nothing of the sort can be seriously claimed. The only intellectual traits about which research has reached some conclusions concern either serious handicaps or some aspects of personality which pertain to psychiatry.

The outcome of this search for precise definitions may seem disappointing, but it is realistic. The concept of "gifts" or of "potential intelligence" is simply not definable; these are empty words that point to no accessible reality. "Developed intelligence" corresponds to the set of all observable characteristics of intellectual activity, but since this activity is multidimensional, any comparison between two individuals can only be arbitrary. Finally, "measured intelligence" makes comparisons possible, but the aspects of intellectual activity which it characterizes are so partial that it seems inappropriate to use the term "intelligence" in reference to it.

The imprint made on the mind of a listener or of a reader by a discourse results less from what is explicitly stated than from what is implicitly assumed or suggested by the words. In referring to certain especially remarkable traits of intellectual activity as "gifts," we reinforce the hypothesis of a quasi-fatal biological determinism of these traits. When we compare the intelligences of two people and assert that "A is more intelligent than B," we are assuming that this trait is measurable by a single number. Honesty consists of recognizing that these are misuses of language, that is, betrayals of trust.

3. *An Adjective: "Genetic"*

After two substantives, let us now round off this review of a few verbal stumbling blocks with an examination of an adjective, the adjective "genetic." What is meant by the statement that a particular trait, a particular character is or is not "genetic"?

What is at issue is the possible link between the trait expressed in an individual and the content of the biological heritage assembled at his conception. But the word "link" is itself very vague; it can designate either a rigorous causal chain or a concomitance measured by what mathematicians call a "correlation." Before continuing, it is important to stress this distinction.

CAUSALITY AND CORRELATION

From the very beginning of a course in statistics, professors warn their students against the logical error which consists of believing that a correlation between two measures indicates a cause-effect relationship between the two phenomena measured. It is easy to give examples which make this error grotesque:

> —Let us plot the monthly curve for coal consumption, then, on the same graph, the curve for mortality among the elderly; the two curves are remarkably parallel—they peak in January-February, drop in June-July. Should we conclude that a reduction in coal consumption would bring about a decrease in the number of deaths among the elderly?
>
> —Let us put two questions to passengers on an underground train in Paris—one concerning the number of days spent annually at winter sports resorts, the other their annual rent bill. On the whole, residents of the fashionable districts who pay high rents spend more time at winter resorts than the tenants of large housing projects in the outlying suburbs. The two variables, annual rent bill and length of winter vacation, are closely correlated. Should one conclude from this that an increase in the rent for subsidized housing would lead to longer winter vacations for working-class people?

Beyond the obvious ridiculousness of these conclusions, let us become aware of the danger of an interpretation whose absurdity may very easily escape us. The logical error is sometimes well camouflaged, but it remains basically the same. For instance, finding that children with an IQ under 120 have difficulties graduating from high school, a child psychiatrist recommends directing them very early on toward shorter streams, which would spare them failure and do away with overcrowding in high schools. The argument here is of the same nature as that concerning restrictions in coal consumption to reduce the number of deaths among the elderly. But there is a danger that it could be taken seriously, because it carefully camouflages the real problem, which is the search for a common cause for the two results between which a correlation was observed.

The "link" between two phenomena may mean either that one is the cause of the other (sunrise and daylight)—this is therefore a *causal link*—or that knowledge of one supplies information about the other (annual rent and length of winter vacation)—this is therefore a *statistical link*.

When we want to act as effectively as possible, a link of the second kind may very well suffice, at least locally: information about rent gives us information about vacations, which may, for instance, allow us to organize an advertising campaign for a brand of skis. With this approach, we do not concern ourselves with the reasons why the parameters under study evolve along parallel lines. Our objective is to use the available data to the best advantage, not to understand. Still, we must ensure that our action does not consist of manipulating one of the phenomena in the hope of modifying the other: in so doing, we would be interfering with the statistical link on which we wish to rely.

On the contrary, when we want to elucidate a mechanism to explain our observations, it is a link of the first kind that has to be sought. Lengthy research and numerous experiments may be required to demonstrate it and specify its modalities. Repetition of the same protocol under identical conditions can alone guarantee us that the first observations were not just a result of chance. To isolate the proper effect of each factor, we have to compare the results obtained under varying conditions of observation and experimentation. At the end of all this work, we will be in possession of an explanatory model where the interplay of the various determinisms will be clearly described; a causal and not just a statistical link will be elucidated.

CAUSALITY AND COMPLEXITY

This causal link may be direct: I press the switch, the bulb lights up; I let go of the stone, it falls. However, quite often the phenomena under study are part of a mechanism so complex that we renounce describing it in full; we merely remark the variation produced in a particular characteristic by the modification of a particular factor. The link thus observed is the result of a tangle of mechanisms, but it does not itself have the rigor of a causal link and may have paradoxical aspects. When my car is at a standstill, I have to engage the clutch in order to begin to move forward; a little later, when my engine is laboring and is about to "stall," I have to disengage the clutch in order to continue. The answer to the question "Is the forward movement in a causal relationship with the fact of engaging or of disengaging the clutch?" obviously depends on the circumstances. The two characteristics which I had arbitrarily isolated, the position of the clutch and the forward movement of the car, are indeed linked by perfectly rigorous mechanisms, but these are so complex that the same clutch position can have opposite effects on the movement of the car.

As soon as a movement is somewhat complex, the notion of "consequence" loses its clarity and even its meaning: the falling of the stone is a consequence of the fact that I let go of it, but it is misleading to present the acceleration of the car as a consequence of having engaged the clutch; this action is only one of the many factors whose interaction leads to the acceleration.

An event is usually the result of a great number of circumstances which interact with each other, no one of which on its own is sufficient, that is, capable on its own of triggering the event independently of the others; it is a "consequence" of the whole and not of any one arbitrarily chosen factor.

However, our mind is ill accustomed to taking interactions into account; for the sake of our intellectual comfort, we grasp at "causes." This need is particularly common when the object being observed is another person, or ourselves: we are eager to know what is the cause of a particular trait manifested by Mr. X. For centuries it was customary to rely on the concept of divine will for an explanation. In the modern era, there is a tendency to rely on genes, but what exactly do we mean when we say that "this trait is genetic"?

IS SCHIZOPHRENIA "GENETIC?"

When the causal chain between the trait being observed and the gene or genes on which it depends appears to be short, the meaning is clear: for each modality of the trait can be found one or more corresponding genetic endowments, and vice versa. Thus, the rhesus blood system is "genetic" because a correspondence between the two observable modalities of this system, "rhesus positive" or "rhesus negative," and the genes present at a certain location (designated by the term *locus*) on chromosome number 1 has been established: these genes belong to two categories, designated by the letters *R* and *r;* individuals with two *r* genes at this locus—the *rr* homozygotes—are "negative"; the others—the *RR* homozygotes and the *Rr* heterozygotes—are "positive."

The same is true of a great number of blood systems, of immunological systems, of metabolic disorders, or of some rather trivial traits (such as the ability to roll one's tongue into the shape of a drainpipe or to taste the bitter chemical phenylthiocarbamide). All these traits are strictly "genetic," because they are a direct consequence of the presence, in an individual's genetic endowment, of particular combinations of genes.

But most often the influence of the genetic heritage is probable, or even obvious, without it being possible to establish a direct causal link.

The study of the transmission of schizophrenia clearly illustrates the difficulties involved. We are not going to enter into the debate about the meaning of the word "schizophrenia" but are assuming, as a working hypothesis, that those in the know, the psychiatrists, are capable of making one of three diagnoses for each individual: "schizophrenic," "schizoid," or "normal." It has long been noted that "schizos" are more frequent in certain families, which points to a possible genetic determinism, but other explanations evoking behavior alone can just as easily be imagined, because behavior too, independently of any genetic factor, is transmittable.

One possible way of pinpointing a suspected genetic mechanism is the analysis of genealogies. The most extensive investigation was no doubt that carried out by a multidisciplinary team including researchers from the CNRS, from the psychiatric unit of Necker Hospital, and from the Institut national d'études démographiques;[1] it studied 25 genealogies, including a total of 1,333 people, for whom a diagnosis in one of the three terms mentioned could be made.

These genealogies were confronted with various models explaining the transmission of these traits through a particular genetic mechanism: for example, the action of genes at one locus, interaction between genes present at two loci, and the cumulative effects of genes dispersed over many loci. Certain models incompatible with the observations could thus be eliminated. From the others, a "best" model was chosen using the classic method of "maximum likelihood." This method involves the calculation of the probability of the genealogical sequences actually observed, assuming that a particular model were to accurately represent the underlying mechanism. The model for which the probability is highest is designated "the best." This conclusion, based on many hypotheses, is of course provisional and cannot in any sense be presented as proved; it merely corresponds to the attitude which fits most closely with the information supplied by the genealogies analyzed.

Oddly enough, the model which proved best for the transmission of the "schizo" trait in the families studied is one of the simplest; it links this trait to the genes present at a single locus. According to this model, individuals endowed with two "normal" genes, the *NN* homozygotes, would never be affected; those with a double dose of a certain *x* gene, the *xx* homozygotes, would manifest schizophrenia with a probability of 25 percent, would be schizoid with a probability of 35 percent, and normal with a probability of 40 percent; finally, the heterozygotes *xN*, endowed with a normal gene and an *x* Gene, would never be schizophrenic, but could, with a slight probability in the order of 5 percent, be schizoid.

Supposing that this model were to be confirmed by further observations, could the *x* gene be presented as the "gene for schizophrenia?" This would be very misleading since three-quarters of those with two copies of this gene and all of those with just one copy would not be schizophrenic.

A simple calculation, obtained by applying the equations expressing the balance of genotypes in a population (the Hardy-Weinberg equations, presented in *In Praise of Difference*, pp. 19–21) shows moreover that the heterozygotes *xN* are much more numerous than one would expect: in Europe, people affected with schizophrenia represent 1 percent of the population (a percentage that is just as valid for Germany or the Soviet Union as for France); the homozygotes for the *x* gene are four times as numerous, since only a quarter of them, according to this model, are affected by the disease. The frequency of the *x* gene in the collective genetic heritage would therefore be about 20 percent (the square root of 4 percent), and that of heterozygotes 2 x 0.2 x 0.8 = 32 percent. In other words, close to one third of Europeans would, without being aware of it, be carriers of the *x* gene.

It would not therefore be correct to present this gene as the "cause" of schizophrenia. People afflicted with this trait would not, in fact, have been the prey of any fatality, since three-quarters of those who had received the same genetic endowment would have escaped the disease thanks to the intervention of a favorable environment. At the very most, one could say that the *x* gene is necessary but not sufficient, or that it predisposes to this trait—which is much more restricted than the assertion that schizophrenia is "genetic."

COLOR, UNEMPLOYMENT, AND GENES; SEX, MATHEMATICS, AND GENES

In order to more clearly illustrate the difficulty of interpreting genealogical studies, let us imagine a Martian, very well informed about the various techniques of population genetics but incapable of distinguishing black from white. On landing in South Africa, he decides to study a trait which strikes him as very important for the fate of individuals, unemployment. First, he notices a very clear link between successive generations of the same family: in certain genealogies, all individuals are free of it; in others they are almost systematically affected by unemployment. He concludes that, very probably, this trait is governed by the genetic heritage. He broadens and refines his observations, imagines genetic models, and tries to find the "best" model with the help, for instance, of techniques for maximizing likelihood. The odds are that he will conclude that the trait "unemployment" is easily explained by the presence at three or four loci of a certain *u* gene. Could unemployment be a "genetic" trait in the human species?

In fact, his research will have led him to discover the *c* genes which give individuals lighter or darker skin, depending on their number in the genetic heritage (it is known that individuals with no *c* gene are white and that the skin becomes darker according as the number of *c* genes present increases). Now color, in the society under study, is strongly correlated with the risk of unemployment; the conclusions drawn by our Martian are therefore perfectly correct; they permit accurate prediction, they are efficacious. But they give no indication of the mechanism at work. Changes in the social rules would make the observed link disappear totally.

The logical error here consists, once again, in studying a phenomenon which results from complex interactions by artificially and arbitrarily isolating one of the factors involved. Our mind is poorly trained for thinking in terms of interactions and tries to replace the reality with models where the various causes act independently. All the questions concerning "the innate and the acquired" are typical of this approach; they do not merit a reply, since they deny the reality which they claim to study.

Unfortunately, the absurdity of these approaches is often hidden by the use of scholarly words or by reliance on complex mathematical formulations. Confronted with a pseudoscience, which is mere pedantry, one is in danger of being fooled by arguments whose ineptitude would be glaringly obvious if they were formulated in ordinary terms.

Newspapers provide, alas, frequent examples of such abuses. Let us consider the recent case of a Parisian evening newspaper (not *Le Monde*) proclaiming with a bold headline, on January 3, 1980, that "having a good head for maths" is linked to "a hereditary gene (sic!) less frequent in women." The author begins with a bow to women, who are not necessarily all stupid (look at Marie Curie or Anne Chopinet), and then announces that "a team of American researchers" have demonstrated that the difference in mathematical ability between men and women "is above all a *genetic* question."

No reference is supplied: what team of researchers? at what university? in what scientific journal did they publish their results? The reader is expected to believe and have confidence, since "American researchers" are involved!

But above all the term "genetic" is used here without any precautions. To be sure, sex is genetically determined; the differences, notably hormonal, between men and women lead to differences in their intellectual behaviors.

However, ascribing "genetic causes" to a trait as subtle and mercurial as mathematical ability would require precautions which are not even mentioned here.

For the adjective "genetic" to have a meaning, it is necessary to define it in a very restrictive way. The best strategy would be to use it only in those cases where every modality of the trait under study corresponds to one or more associations of genes. Using it in reference to traits which seem more or less dependent on the genetic heritage can only be a source of confusion. In reality, any trait, no matter what it is, depends on this heritage, since it is expressed and developed in an individual who could never have existed were it not for his genes. Let us be especially wary of the connotations of fatality or even malediction associated with this adjective, when, usually, only a vague predisposition is involved.

QUESTIONS

Major advances in knowledge result less from the discovery of a reply to an existing question than from the formulation of a new question or, more frequently, from the new formulation of an old question. When the thinking of researchers revolves in futile circles, constantly blocked by insurmountable paradoxes or by a growing mass of complications, the solution is generally made possible by someone who poses the problem in new terms.

The explanation of the movement of the planets by means of circles centered on the sun or of circles whose center itself travels a heliocentric circle was becoming, at the end of the fifteenth century, more and more complicated according as increasingly precise observations became available. By replacing the circle with the ellipse, Kepler made most of the problems considered paramount by astronomers irrelevant. The central concern ceased to be the interlocking of circular movements, and was replaced by that of specifying the characteristics of the ellipses.

Similarly, in the nineteenth century, the explanation of the transmission of traits from parents to children was locked in contradiction. If, as seems natural, each of a child's traits results from the fusion of the paternal and maternal contributions, this trait will be close to the arithmetical average between the traits of the two parents; the repetition of this process at each generation should systematically lead to a gradual homogenization of the population; however, no such homogenization actually occurs—on the contrary, diversity seems everywhere to be maintained. There is a glaring incompatibility between the most natural explanation and the observed reality. The solution was supplied by Mendel's hypothesis: parents do not transmit their "traits" but half of the "factors" (we now say the "genes") which govern these traits; for each elementary trait, an individual thus possesses two factors, one of which comes from his father, the other from his mother; they coexist without coalescing, remain unaltered throughout existence, and are in their turn transmitted, unchanged, to the next generation. This hypothesis, which we now know to be in complete agreement with the biological reality, is contrary to common sense, because it replaces the apparent oneness of each being with an organism under dual control. This systematic copiloting is difficult for our mind to accept; even today, more than a century after Mendel's discovery, many arguments are implicitly based on the transmission of traits and not of genes.

The most common kind of laziness does not consist of refusing to work but of refusing to summon our imagination in order to find answers to the questions that are asked of us. We are in general prepared to carry out lengthy calculations, to solve complex equations, and to develop intricate and laborious arguments, but our mind stalls at the prospect of seeking and forging questions formulated in new terms. The mathematician T. Guilbaud likes to claim that we are as old, not as our arteries, but as our algebras, that is, as our capacity for changing from one day to the next the models by which we represent the real world. A daily exercise in this domain can be as rejuvenating as a jogging session (see box 2).

Out of laziness we expect science to answer our questions; scientists themselves sometimes play along with this and do not protest at being introduced as "those who know," those who have answers. This is sometimes true, but science is a territory which is defined above all by its frontiers, and on the frontiers of science, everything is in question. The most exciting areas of research are those which are teeming with questions that are as yet but partially formulated.

The way in which the scientific enterprise is presented to the public leads to considerable distortion. News reports always deal with the breakthroughs, the successes, the answers. As soon as a new satellite of Saturn is dis-

Box 2
Intellectual Jogging

As recreation, let us consider the following logical schema proposed by the English mathematician Bertrand Russell: I go to the university libraries and for each one draw up a catalog of all the works it contains; having completed this catalog, before putting it on the shelf I may find it appropriate to include the catalog itself in this catalog, since it constitutes a new item in the library. I can also decide not to make this addition. I am thus led to realize that two kinds of catalog exist, those which contain themselves—they form the set A—and those which do not contain themselves—they form the set B. Considering the various elements of this set B, I can then draw up the catalog X of the catalogs that do not contain themselves; to which set, A or B, does this last catalog belong?

—to A? Then it contains itself. Therefore, I must register X in X, but X gives the list of catalogs that do not contain themselves—therefore X belongs to B.
—to B? Then it does not contain itself. X does not feature in X—it therefore does not belong to B.

It is strange to find that this paradox, which is related to the axiomatic difficulties to which we referred on page 00, is at first poorly understood by minds that are unprepared. A rather strenuous effort is required to follow this discourse, which is actually very simple. After which, this logical development seems obvious, and the greatest difficulty then is to understand why others have difficulty understanding it!

covered, as soon as a new particle has left a trace of its passage in a bubble chamber, a bulletin carried by all newspapers informs us of it. But no special report announces a change in the formulation of a question that long seemed resolved. Questions, which are at the very heart of the dynamics of research, do not constitute events; they therefore spread very slowly beyond the small circle of researchers involved. This noninformation, which is due to difficulties proper to this diffusion and not to the ill will or incompetence of those providing the information, can have serious consequences; it allows certain people to present their personal ideological options as being the only ones in accordance with the "lessons of modern science" or as being "scientifically demonstrated." In order to combat this abuse of science, which may lead to the perverse use of certain scientific results (as in the case of the renewed attempts at promoting eugenics), it is necessary to take stock of the current questions without excessive defeatism but without triumphalism.

In each discipline scientists should resolve to make a wide public aware of the conceptual difficulties besetting them; in this way, they would give everyone the arms necessary to combat the misinterpretations made of discussions between scientists. As examples, we will explore in the chapters that follow three domains where biology provides more questions than answers, but also makes it possible to formulate these questions with a clearer meaning than hitherto: education, social organization, and the evolution of living things.

5.
Biology and Education: Intelligence, Its Support, and Its Development

Educating: out of a helpless infant making a grown up person, out of an individual whom biological processes have implacably formed since the chance encounter of a sperm and an egg, making a being whose place in the unfolding of human history is unique. Educating is "making," but out of what material, toward what end, and subject to what constraints?

This debate is so important that it is difficult for it to escape the contamination of ideologies or of political options. This is because it is a question, quite simply, of our freedom.

Taking up arguments developed especially by Calvin, the Dutch bishop Jansenius shook the seventeenth-century Church by adopting a seemingly irrefutable viewpoint: God knows all things; therefore He knows whether or not my soul will be saved; my salvation is already decided—it was so even before my birth. According to Jansenius, I am *predestined.*

A century later, in 1773, in a famous master's thesis presented to the Academy of Sciences to which we have already alluded, Laplace showed that the laws governing the physical world are rigorous and that "the present state of the system of Nature is a result of what it was an instant before; a perfect knowledge of the present would therefore allow us to determine with certainty and precision the state of this system at every moment of the past and of the future"; according to Laplace, the universe is *predetermined.*

A certain way of presenting results obtained since the beginning of the century by genetics, or recent research carried out notably by American zoologist Wilson and grouped under the title "sociobiology," sometimes attempts to persuade us that our behavior, including our intellectual traits, is governed by rigorous mechanisms encoded in our genetic heritage; according to Wilson, we are *programmed.*

To be sure, it is no longer a question of the salvation of our soul or of the future of the universe, but of our "success" in society; it is no longer a question of God, but of our genes. However, the debate remains fundamentally the same: the direction our life has taken, the being that we have become, are they the result of a preestablished program or of the experiences whch we have, of course, partly endured, but also, to some extent, controlled? Is our destiny preordained, or can we shape it? Does education reveal only our innate capacities, or does it mould us to the point of transforming our intellectual development?

To these questions, science, and especially biology, provides some partial answers; here again, it is necessary to eliminate the pseudoscientific statements of certain doctrinarians who lace their statements with a vaguely mathematical jargon.

ONTOGENESIS AND EPIGENESIS

Our entire organism participates in our intellectual activity; however, a privileged role is played by an especially complex structure, the central nervous system. Its active elements are essentially specialized cells, neurons. As early as the third week of embryonic life, a plate, called a "neural plate," appears, which gradually lengthens and whose sides come together to form the neural tube; the layer of cells lining the inside of this tube becomes differentiated and increases in number, gradually giving rise to the nervous system in all its complexity. From the second month of fetal life, its activity can be observed.

At birth, the 50 or 100 billion or so nerve cells which the individual will have at his disposition throughout his life are already in place, but they have not reached their definitive size and, above all, they are not surrounded by the insulating sheath which will make them functional. The weight of the brain is then only about 350 grams; its growth is very rapid at first because of the "myelinization" of the nerve fibers, that is, their being insulated by fatty layers; then this growth slows down until the maximum weight, 1,300 to 1,400 grams, is reached at puberty, after which a slow decrease begins, and at 75 years of age, this weight is 10 percent less than its maximum. Every day, in fact, about 50,000 neurons are put out of use (that is, 1 billion in sixty years, which, in reality, is but a very small proportion of the total).

The role of each neuron is to receive and transmit information coded in the form of electric impulses; to carry out these transfers, it is connected to other neurons through the intermediary of liaison structures called synapses, whose number, variable depending on the role of the neuron, can exceed 20,000.

Let us attempt to grasp this system more clearly: the total number of our neurons is equivalent to ten or twenty times the total number of people on earth (think of the countless throngs living in Asian or American cities); each of these neurons is in permanent communication with several thousand others; this is the unimaginably rich network that we have available to us for perceiving our environment, for thinking, and for acting. (Yes, this richness is really unimaginable:

let us transpose the myth of Dr. Faustus and assume that a French person were to sell to the devil, not his soul, but his entire network of synapses at the more than reasonable price—givén the precision required to make this item—of one franc each; his riches would be such that he could, on his own, pay all the direct and indirect taxes of his compatriots for two centuries.)

The development of this hypercomplex network is obviously dependent on our genetic heritage; it is, of necessity, the genes which supply the recipes for making the proteins involved in the composition of the various elements of this system or which regulate its functioning. But is it conceivable that its very structure is genetically programmed?

The order of magnitude of the number of genes involved is in the hundred thousands; that of the number of synapses is in the order of a hundred thousand billion; the former could specify the latter in a rigorous manner only if they were elements of a very simple structure, which is obviously not the case.

It seems still more difficult to imagine a precise determination of the structure of the central nervous system by the genetic information when one recalls how this system develops from the initial egg. From before birth, as we have seen, the individual's supply of neurons is complete; now the duration of intrauterine life, nine months, represents 400,000 minutes: the child, during this period, manufactures an average of 250,000 per minute, a rhythm which can no doubt reach 500,000 or even more at certain phases of his development. It is difficult to see how complex structures could be finalized at such frantic speed under the strict control of the genetic heritage.

Biologists J.-P. Changeux and A. Danchin[1] and mathematician P. Courrège proposed a way out of this paradox, using the concept of *epigenesis*. The genetic program no longer corresponds to the rigorous definition of immutable sequences, but to variable realizations of which only the average modality and the dispersion around this average are determined by this program. In other words, the unfolding of the program leaves an essential role to *chance*.

This might be the case in particular with the development of the synapses; they are defined by the corresponding genes only in a vague, unfixed way; they then become stabilized in a specific way as a function of stimuli coming from outside: "The epigenesis of the nervous network corresponds to the transformation of temporal elements, provided by chance events in the environment, into a geometrical organization. The environment draws an imprint on the genetic envelope."[2]

One illustration of this process comes from the recent discoveries of Japanese biologist Tadanobu Tsunoda concerning the localization of certain reflexes.[3] It has long been known that the two cerebral hemispheres have different functions: the left hemisphere is responsible for language and for the processes of logical thought, while the right governs nonverbal functions. This separation seems universal, regardless of the cultures involved.

Professor Tsunoda set out to specify the localization, either on the right or on the left, of reflexes linked to sounds received at varying intensities in the two ears. In the course of numerous observations carried out on his compatriots in Japan, he noticed that for certain sounds the right ear was "dominant," which corresponds to a localization in the left hemisphere of the brain.

He then studied the reflexes of non-Japanese and found that this result was reversed: among Europeans, Africans, or Chinese, it is the left ear which is dominant under the same conditions; only speakers of a Polynesian language react like the Japanese.

But the most surprising result is that Japanese living in the United States and whose native language is English react like Westerners, while the few Westerners whose families have immigrated to Japan and speak its language react like Japanese. Professor Tsunoda concludes that the localization of this reflex is governed not by genetic factors but by the characteristics of the linguistic environment (the great abundance of vowels in the Japanese language seems to him to be a specific trait that is adequate to explain the difference observed in his fellow citizens). For him, "an individual's native tongue is closely linked to the development of the mechanisms governing emotion in the brain."

That is a very specific example of brain structuring as a function not of the genetic material, but of the individual's experiences.

THE ROLE OF CHANCE

It has long been noted that the successive stages in the development of the embryo evoke, in a fairly precise way, the various species that succeeded one another in the course of evolution before culminating in man. This somewhat approximate observation is summed up by the classic statement "Ontogenesis recapitulates philogenesis."

It is significant that, by two totally different routes, the role of chance should be recognized simultaneously both in ontogenesis and in philogenesis. Indeed, the evolutionary models that have been developed by Neo-Darwinism over the past three-quarters of a century are, as I will explain in chapter 7, challenged by observations on the structure of proteins showing an unexpectedly large amount of polymorphism. To resolve this difficulty, certain researchers stress the role of chance in the transformation of genetic structures over the generations. The transformation of species is not only a result of the determinisms imposed by natural selection; it also results from purely aleatory mechanisms (it might be added that sexual reproduction is one of the most efficacious means of introducing chance).

Similarly, theories explaining embryonic and fetal development as the unfolding of an inexorable genetic program are challenged by evidence of a considerably greater richness of information in the finished product than in the formula for production; one way out of this paradox is to attribute a decisive role to aleatory changes.

In both cases, reality is seen as the result of the choice that chance makes among the set of possibilities: determinisms play an important role because it is they that limit the field of possibilities, but it is chance that has the last word.

This vision can be likened to the reflections of certain biologists and computer scientists on the "creation of information from noise" and can be illustrated by the principle governing the "perceptron," a machine capable of learning to recognize forms.[4] It can, for example, automatically recognize the letter *b* in spite of the many variations in the way that it is written by different people; beyond the diversity of forms, it is necessary to be able to perceive the structure that causes a particular form to be identified as being, or not being, the letter *b*. This problem can be solved by means of the electronic calculators with which we are all familiar, capable of making logical decisions with fantastic speed, but the necessary programs are terribly complex. One can adopt an entirely different approach which requires neither calculations nor logical decisions of the machine; this apparatus is endowed with organs of perception (a few hundred photoelectric cells) linked to processing organs (a few hundred amplificators), themselves connected to an organ for response, emitting a + signal or a − signal depending on the total intensity of the currents received.

The essential particularity of this network of wires called a "perceptron" is that the connections and initial settings are carried out by chance: each organ for perception is linked to about twenty amplificators chosen in an aleatory way, and these amplificators are set at any amplification coefficient. The machine leaving the factory therefore does not know how to do anything, since it was made without a plan and without precise instructions; its structure is, in part, aleatory. But it is capable of being trained. To do this, the letter *b* is shown to it, and the setting of the amplificators is modified until a + signal is emitted. Experience shows that, after about forty attempts, the reply given is accurate with a probability close to 100 percent.

It would, of course, be very simplistic and misleading to claim that the functioning of our brain is similar to that of a perceptron or of similar machines. The complexity of the networks of the central nervous system is of an altogether different order from that of the few thousand components and conducting wires of these apparatuses. However, the analogy is still appropriate. What characterizes these apparatuses is both their contingent structure and the superabundance and redundancy of their organs (such a great number of wires and contacts to distinguish the letter *b* may seem entirely out of proportion). It is precisely this conjunction of redundancy and contingency that gives them their essential capacity, the capacity for learning. The central nervous system with its thousands of billions of contacts is no doubt superabundant; it probably results from a process that is, in part, contingent, but it is because of this very combination of redundancy and chance that it too is capable of learning.

Moreover, reflecting on the functioning of these machines enables us to better understand the impossibility of replying to the question so often asked, and which underlies the essence of all thinking developed in this domain: of the characteristics of the intellectual tool, what proportion is innate, what proportion acquired?

THE ROLE OF THE INNATE AND THE ROLE OF THE ACQUIRED

Let us get back to our perceptron: if the conducting wires that unite the organs are of poor quality and do not allow the electric current to pass, the apparatus will never work—likewise, if the worker who constructs it does faulty welding. The way it performs therefore depends on the conditions under which it was made, let us say on what in it is "innate." But, as we have seen, this apparatus can do nothing until a series of adjustments have created a certain structure which allows it to recognize, almost without fail, a particular form—for example, a specific letter—in spite of the variable forms that this letter can take; the way it performs is therefore a function of its successive experiences, let us say, of what in it is "acquired."

The application of this analysis to cerebral functioning is clear: our intelligence results both from an initial blueprint and from our training; it results both from something innate and from something acquired. It seems natural to then ask how much each of these contributes. We saw in chapter 2, when dealing with the pitfalls attending addition, how this type of breakdown, toward which our mind tends as though naturally, leads to an impasse.

It is very important to reflect on the meaning of this question—important too, once we have discovered that it can have no meaning, to examine the motives that led us to ask it.

Let us recall the essential result: attempts at establishing the proportion to which various factors contribute to a trait can have meaning only if these factors (in this case, innate and acquired ones) have additive effects on intellectual activity. If this additivity is not proved, interaction effects have to be taken into account. The analysis of the dispersion of the parameter under study (for example, IQ) does not then include just two terms, dispersion due to genetic factors and dispersion due to environmental factors, but three terms, the extra term representing the dispersion due to the interaction between these factors.

Naturally, no one questions the existence of this interaction: the effect of the genes depends on the environment, the effect of the environment depends on the genes.

The roles of the innate and the acquired in the expression of a trait can be compared to those of grammar and vocabulary in the meaning of a sentence. "The cat eats the mouse" has meaning only if I understand the words "cat," "mouse" and "eat" and if I know the rule which stipulates that the substantive preceding the verb represents the doer of the action and that following it the object of the action. The rule without the words is mute, the words without the rule have no impact. Who would think of measuring the relative importance of each of them?

Likewise, isolated genes are mute; the input of the environment without the genes has no effect.

In all logic therefore, the problem of "the innate and the acquired" should no longer be raised. However, this is a domain where dogmas are infinitely more powerful than logic; we must be prepared to continue to frequently read peremptory statements specifying the percentage by which genes determine intelligence, 80 percent being the number most frequntly quoted.

It would be relatively easy to argue about such statements if the number proposed was simply incorrect; if the correct number were 30 or 90 percent, agreement would be reached in the end. But this number is not incorrect—it is absurd.

If someone told me that the moon is 500,000 kilometers from the earth, I would say to him that this figure seems wrong; we could both agree to check it in reliable sources and to accept the figure indicated by one of them. However, if he said that the moon was 10,000 tons away from the earth, I could only voice my disagreement without being able to suggest another number. His statement is no longer inaccurate—it is nonsense. Experience proves that it is unfortunately much more difficult to combat nonsense than error.

INTELLIGENCE, ITS MEASUREMENT, ITS HERITABILITY

We explained in chapter 4 how psychologists, eager to give their discipline the scientific aura bestowed by number, have attempted to translate intellectual performances by means of markers on various scales, and to synthesize them by means of a parameter irritatingly called "intellectual quotient."

The scientist's first surprise, expressed notably by Francois Jacob, is that a characteristic as multiform as intelligence can be reduced to a single measure.[5] This word "intelligence" is used in everyday speech with very diverse meanings. Several columns of the Robert dictionary are devoted to it;[6] qualities as diverse as perspicacity, discernment, reflection, creativity, and sensitivity are in different ways, varying with each individual, implicated in intelligence. Before continuing with this discussion, some precautions are therefore necessary.

One of the dangers of this parameter is that, since it is presented as a number, one is tempted to treat it as such and to use it in various arithmetical operations when, in most cases, these operations have no meaning.

To be sure, we commonly use other numbers of ambiguous meaning—for instance, for temperature—but we were taught, along with their definition, the techniques for their use. We know, for example, that the average of the temperatures 30° and 50° has meaning only when account is taken of the specific mass and heat of the two bodies put in contact; the average of the numbers 30 and 50 has in itself no meaning. On the contrary, the temperature obtained by mixing 100 grams of water at 30° and 200 grams of oil at 50° (whose specific heat is represented by σ_h) is a perfectly defined number, which can be calculated by the formula

$$t = \frac{(100 \times 30) + (200 \times 50 \times \sigma_h)}{(100) + (200 \times \sigma_h)}.$$

A number can be used in a pertinent arithmetical operation only if its significance is defined—in other words, if it measures a parameter linked to precise properties of the world around us. Such is indeed the case with temperature; but in what formula can one validly use IQ? Even the average of two IQs can have no meaning; one can then ask oneself if it is appropriate to represent it by means of a number.

By very subtle but necessarily arbitrary procedures, one can calculate a number labeled IQ; this number can measure nothing other than the characteristics being observed, that is, the performances achieved in certain tests. Its only value comes from its empirically observed correlation with other parameters measured on the individual concerned. Binet's initial objective was to evaluate the risks of failure at school; he therefore imagined tests and weighted the marks obtained on these tests in such a way as to obtain a number with the best possible correlation with the individual's level of success at school (moreover, he did not use the term "intellectual quotient," which did not appear until later). This measure can therefore rightly be considered an indicator of potential for success at school. To extend its meaning to include intelligence, it must be assumed that the latter is represented by performance at school, which is not absurd, but considerably restricts the meaning of this word.

The most honest position seems to be to assume that we do not know exactly what entity the number labeled IQ measures, nor even whether this entity is definable. However, it has been proved empirically that it is correlated with success at school and, consequently, in a society like ours, it is also correlated with social success. We are roughly in the same position as someone observing a frog in a jar; he does not know why the frog goes up or down the ladder, but he knows from having observed it over a period of time that the higher the bar on which it stops to rest, the greater the chances of fine weather the next day.

There is a willingness to expend considerable effort on the calculation of an IQ because this IQ will be used as a basis for certain decisions. Precautions are obviously necessary, which are rarely remembered; since it involves a measure, it is appropriate first of all to establish its degree of stability and accuracy.

Strangely, psychologists seldom address this question of accuracy. Let us quote the number proposed by P. Dague[7] and confirmed by M. Carlier:[8] according to them, the range of dispersion "at 95 percent" is about ± 10 for IQs above 85, ± 5 for IQs below 85. In other words, stating that "this child has an IQ equal to 108" means that "IQ measurements carried out for this child would, 95 times out of 100, lead to a result between 98 and 118, 5 times out of 100 to a result outside this interval." Certain psychologists consider that these authors are rather pessimistic and assume an interval of ± 7. This imprecision of the margin of imprecision of IQ is in itself indicative of the vagueness of the object being evoked. One can indeed rigorously define a "confidence interval" for the measure of a mass or length, because there is an actual, but unknown, mass or length that one is trying to situate between two extreme values; on the contrary, there is no "true IQ" independent of the measure that one calculates. Be that as it may, let us remember that every mention of an IQ should systematically be accompanied by an indication of its precision. The deontology code forbids physicians certain actions or certain imprudent words; would that it forbade saying to parents "Your child's IQ is about 97," when the only thing that one can rightly say is "Your child's IQ has 95 chances out of 100 of being between 87 and 107 (or between 90 and 104)!"

As for the stability of IQ, it is even less well understood than its accuracy. One result that seems to me to be particularly enlightening is provided by a study carried out for many years by INED on the intellectual level of school-age children. Note that these studies have shown an average increase of 10 points in four years among immigrant children taken as a whole. This global evolution that was found for all the samples studied, regardless of country of origin, is the result of individual evolutions, some of which were much more rapid. The results of the INSERM study of adopted children, which will be presented farther on, also show how the evolution of a child's IQ depends on his environment.

The essential point is that an individual's IQ is not a permanent trait like his blood group or his sex. It is a rather imprecise measure, which corresponds to the current state of some aspects of his intellectual abilities. Let us above all avoid seeing it as a label to be attached to each person, definitively marking his destiny.

However, IQ being a number, countless attempts have been made to calculate what is called its "heritability." This scientific sounding word is one of those that has most distorted dialogue between psychologists and geneticists; a chameleon-like word which changes its meaning depending on the discourse, it corresponds in fact to three perfectly distinct concepts:

1. the similarity between children and parents,
2. the proportion of the total variance of a trait that is ascribable to the global effect of all the genes present,
3. the proportion of this total variance that is ascribable to the individual effects of the genes.

Unfortunately, heritabilities 2 and 3 can only be defined by means of completely unrealistic hypotheses about additivity; in the absence of these hypotheses, the analyses of variance are, as we saw previously, meaningless. As for heritability 1, it measures only a correlation and does not in any sense allow one to specify the determinisms involved. Techniques, often complex ones, have been developed to determine these three heritabilities; the necessary calculations can always be carried out, but neither their difficulty nor their accuracy give meaning to the final result, while the lack of realism of the hypotheses does not even allow one to define the meaning of the parameters evaluated.

Once again, an extreme case of the type of misinterpretations that result from the confusing of the three concepts denoted by the word "heritability" is provided by the writings of the French child psychiatrist whom we have already quoted. His books are especially useful to teachers who are looking for examples of mistakes to be avoided; they contain numerous examples of extreme cases which are certain to capture the attention of students. For him, "the contribution of heredity to intelligence is about 80 percent, that of the environment, *what is left,* is 20 percent";[9] there could not be a clearer or more ingenious proof that this argument is based entirely on the addition of the effects of the genes and of the environment, which no sensible person would dare to claim.

GENES AND INTELLECTUAL DEFECTS

In stressing the impossibility of calculating the exact contribution of the innate to the expression of a trait as complex as intelligence, we do not mean to imply that the influence of genetic factors must be denied. It is very clear that all the organs required for intellectual activity are dependent on the genes, both for their development during ontogenesis and for their maintenance and regulation; it is therefore clear that their functioning is "genetic." However, this adjective is much too vague to be used in a discourse that is claiming to be scientific; we dwelt at length on this difficulty in chapter 4. Let us assume, with geneticists, that a trait can be labeled "genetic" if a link has been established between its various modalities and the presence of certain genes in the heritage of individuals. Thus, the rhesus blood system is "genetic," because the "minus" trait is conditional on the presence of two copies of a certain gene designated by the letter *r*.

With a definition such as this, it is clear that intelligence is not itself "genetic," so numerous are its modalities; we can, at most, hope to specify the genetic mechanism for certain traits linked to intelligence. Such is the case for certain forms of mental retardation.

It is known, for example, that phenylketonuria or the amaurotic idiocies (especially Tay-Sachs disease, frequent in certain Jewish groups) are caused by the presence of two copies of a certain gene: these diseases are transmitted exactly like the green-colored cotyledons of the peas studied by Mendel.

The activity of these genes is such that the brain is destroyed or cannot function; IQ is therefore modified by their presence. However, it must be noted that those genes whose effect on IQ has been specified all have a negative effect; there are genes for retardation, but we know of no genes for intelligence.

It was even possible to attempt to count these unfavorable genes, or rather to count the loci that they occupy (a *locus* being the segment of a chromosome occupied by the genes governing a given trait; a locus is therefore the unit of hereditary transmission). To do this, the effect of relatedness of father and mother on the IQ of the children was observed. For example, when the father and mother are first cousins, the child can receive from each of them not two distinct genes but two copies of the same gene that was initially carried by one of their common grandparents; the child is then necessarily homozygous. The probability of this event is called the child's "coefficient of kinship." Most studies show that the IQ of consanguineous children is slightly lower than that of nonconsanguineous children: for children born of first cousins, the difference is about 4 points lower according to a study carried out by Schull and Neel in Japan and about 2.5 points lower according to a study carried out by Slatis in the United States. Results such as these must be used with circumspection, so difficult is it to isolate the "consanguinity" factor from other factors to which it is linked, in particular the socioeconomic status of the family; cousin couples do not, without any doubt, constitute a representative sample of all couples, and one can never be sure of completely eliminating this bias. The decrease observed, 2.5 or 4 percent, must therefore be interpreted with prudence (note that it is so small that cousin couples have no reason to worry unduly about the health of their offspring). To the extent that it corresponds to a real genetic effect, two consequences can be drawn from it.

First of all, these results suggest the existence of a large number of loci where recessive genes with an unfavorable effect on IQ can meet. Arguments whose speculative and even somewhat unrealistic nature must be underlined have enabled N. Morton to estimate at "more than 300" the number of these loci.[10] In spite of our reservations about accepting this result, note that it is coherent with the idea that cerebral functioning can be upset by any number of metabolic disorders, themselves resulting from the presence of certain genes. Moreover, it is not necessarily a question of genes acting directly on the central nervous system. Our organism is an integrated whole involving complex interdependencies; an activity as elaborated as intellectual activity is subject to the most diverse influences.

However, the most important factor brought to light by these observations is the unfavorable effect on IQ of "homozygosity," that is, of the presence of loci occupied by two genes with the same structure. It is not so much the presence of unfavorable genes that lowers the level of IQ as the lesser diversity of the genes present; the adjectives "favorable" or "unfavorable" could not therefore be applied to particular genes, but to particular combinations of genes.

This fact totally contradicts the frequently expressed ideas on the necessity for a eugenic policy. Since Galton and the theories he developed in his famous work *Hereditary Genius* published in 1869, ten years after his cousin Darwin's *Origin of Species,* it is fairly commonly accepted that certain human lines are better endowed intellectually than others. Translating this finding into genetic terms, it is concluded that certain families carry genes that are more favorable and that it would be good for our species if these genes were to become more widespread. How many books and articles have been written to warn against the intellectual deterioration that threatens us if the least "gifted" families continue to be more prolific than the most gifted ones! It is in this spirit that a 1924 Immigration Act restricted entry to the United States for certain human groups; Brigham, the psychologist advising the committee responsible for preparing this law, denounced "the decline in intelligence caused by the immigration of blacks and of Alpine and Mediterranean races" and advocated a restrictive policy based on "science."

As soon as it is accepted that "quality" is linked, not to the presence of a particular gene, but to the presence of various genes, all these arguments crumble. Consider the robust corn in Beauce and in Brie; its strength is not due to its having a particular gene *a* or *b,* but to the fact that, at many loci, it possesses an *a* and a *b* simultaneously, that is, that it is heterozygous *ab*. The so called pure lines, the homozygotes *aa* and *bb,* are both less robust. The question "Is the *a* gene better than the *b*?" therefore has no meaning; quality does not depend on the nature of the genes, but on their diversity. If the "corn" population was being governed by a dictator obsessed simultaneously with the improvement and the purity of the race, the worst catastrophes would be likely, since these two objectives are opposed.

In view of these facts, any eugenic measure intended to increase the intellectual potential of a population is, at best, silly. In spite of this, we constantly come across articles aimed at a wide public where statements based on this nonsense appear—for instance, the articles in a certain magazine praising the initiative taken by a Californian businessman, the founder of a sperm bank to be supplied by Nobel prizewinners!

INTELLIGENCE AND FATALITY

Morton's conclusion, which we said had to be interpreted with great prudence, may mean that diversity of the genetic heritage benefits intelligence, or at least the intelligence quotient; here then is proof that the latter is "genetic"!

To many authors, this word evokes a fatality: intellectual destiny is definitively fixed as soon as the collection of genes is constituted at conception; some among us may be lucky enough to have been well endowed, others may have been losers in this lottery! Here again, the same child psychiatrist whose peremptory statements have been so useful to us illustrates this mechanistic vision very well. He has no qualms about asserting, having measured a child's IQ, that this child

> —will not normally be capable of pursuing third-level education if his IQ is below 120;
> —will not be able to accede to high-level mathematics if it is below 130.

Once again, we are confronted with the error, which we denounced on page 94, that consists of taking a correlation as proof of causality; the English psychologist Eysenck, our child psychiatrist's source of inspiration, commits exactly the same error when he writes:[11] "IQ remains, whether we like it or not, the most important *factor* in success at school"; this argument is equivalent to that which states that the increase in coal consumption is a *factor* in the mortality of the elderly. It is not a question of a *factor,* but of concomitance; it is remarkable that such serious errors of logic should be made in books claiming to be scientific and where there are so many references to "the latest discoveries of modern science."

Statements such as these, without any serious basis, make one shudder. Treating them with contempt and amusement is not enough; we must react, in the name of rigor, in the name also of the future of our children threatened, in the words of American scientists, with being victims of the "IQ archipelago."

In reality, this question about "the innate and the acquired," which we saw to be meaningless except in very special cases, is asked with a view to reinforcing a certain ideological standpoint: it is a question of "scientifically" demonstrating the inevitability of individual success or failure and of thus founding the social hierarchy on natural causes.

When analyzing the meaning of the adjective "genetic," we stressed how limited the gene-fatality association actually is; there is a direct link in the case of the structure of a particular protein or of the specific group expressed for a particular blood group system; the link is no longer direct when one is studying a "trait," even if its determinism seems simple. "Phenylketonuria," a disease due to the presence of two copies of a certain gene, was fatal as long as the metabolic malfunction underlying it was not understood; for the past twenty years or so, a change in the diet, that is, in the "environment," has been making it possible to cure children with this genotype: affliction with this disease which, in the past, depended 100 percent on the genetic heritage now depends 100 percent on the environment.

This triumph over fatality, possible even for a trait whose determinism results from a very short causal chain, is even more possible for complex traits such as the various aspects of intellectual activity. In order to understand the mechanisms involved, it is important to design and carry out the most rigorous studies possible, with a view to specifying to what extent change of environment can change individual destiny. Two main routes can be followed.

STUDIES OF TWINS REARED APART

The first is the study of monozygotic twins reared apart: they have the same genetic heritage; the differences found between their behaviors can therefore be caused only by environmental influences. In order to be able to ascribe their similarities to biological factors, the environments in which they are raised should be entirely different, which is rarely the case. Most studies of this kind were carried out by English psychologist Cyril Burt, but the true facts emerged a few years after his death at the age of eighty-one, in 1971: in essence, the data presented in his many articles were purely and simply invented. Convinced of the genetic determinism both of IQ and of social success, he fabricated observations on a total of 53 pairs of twins. Certain psychologists, to save the essence of Burt's conclusions, have suggested that only his last studies were suspect and that his great age was responsible for the errors made at the end of his career. A close examination of one of Burt's major works on "distribution of intelligence according to occupational class" led D. Dorfman of Iowa University to conclude that in this work also "he had, without any doubt, invented his data."[12] This author was even able to reconstruct the method used by Burt to ensure that the statistical treatment of the pseudoobservations reported in his article would yield the results he wished to obtain. Now this study, which was supposed to be based on observations of 40,000 pairs of children and parents, has been used in countless works and is cited in all the treatises. Thus repeated and reinforced, it comes to be presented (possibly in good faith) as an irrefutable scientific fact, while it is but a swindle.

For all that, studies of twins reared separately must not be abandoned; scrupulously carried out, they may provide precious information.

A University of Minneapolis team began research (some years ago) on adult twins separated from the first weeks of their life. Twenty or so specialists from various disciplines are involved in this study. Given the precautions taken, the results will be rigorous, but will not be available for a long time to come: the current rate is about ten pairs of twins studied per year. At the congress on twinship held in Jerusalem in June 1980, this team presented initial findings from studies of 14 pairs; they bring to light major gesticulatory similarities.

A second possible method of studying the effect of the environment on each person's destiny is to study children adopted by families with a very different social status from that of their biological family. Surprisingly, while the usefulness of such studies was underlined as early as 1913 by the English psychologist Richardson, none was completed prior to that whose results have just been published by a team from the National Institute for Health and Medical Research.

INSERM'S STUDY OF ADOPTED CHILDREN

This study was carried out by a multidisciplinary team from the Institute for Medical Research including researchers, psychologists, doctors, and geneticists.[13] Observed throughout their schooling were 35 children born to families in what is considered to be the lowest socio-occupational category and adopted, at about four months, into families considered to belong to the highest category. Their results were compared with those of their 39 brothers and sisters raised in their natural family, and also to the average results observed in the various socio-occupational classes. The conclusions are remarkably clear: the 35 adopted children have, as a group, results that are exactly similar to the average of the higher category, that in which they were raised; their performances are very much better than those of their brothers and sisters raised in the lower category. Let us look at some of the numbers: by the age of ten, 5 of these adopted children had experienced various setbacks at school (repeating grades) but only 1 pair of twins had had a serious failure (assignment to a "parallel class"); among their 39 brothers and sisters raised by their biological families, 24 children were found to have experienced failure, and, in 12 cases, serious failure.

It is difficult not to take these results seriously: they show quite clearly that access to our society is refused to almost a third of children from underprivileged categories, not because of a biological inferiority against which we are powerless, but because of the environment in which they find themselves. This is not a demagogical statement—it is a simple statement of fact. It gives at least some idea of the changes still to be made by our society if it really wants to provide "equal opportunity."

The results of the classical intelligence tests confirm, as could have been predicted, the facts supplied by the school failure-rates. Let us consider these results, while bearing in mind the imprecise meaning of the averages obtained: the average IQ of the 35 adopted children is about 108.7, that of their brothers and sisters raised by their biological family about 94.6. Remember that the INED study of more than 100,000 children found that the children of "senior executives" had an average IQ of 108.9 and those of "workers with few qualifications" an average of 94.8. The near perfect coincidence must not delude us: let us conclude, however, that the IQs of the adopted children observed in this study are perfectly similar to those of children born and raised in the same environment.

To be sure, the numbers studied are relatively limited, but extensive methodological precautions were meticulously followed, which gives precise meaning to the results. The lesson to be drawn from the results is clear: social heredity, that is, the transmission of social status from parents to children, depends essentially on the environment, which has effects that are sufficiently powerful to hide the possible consequences of genetic differences.

BEING OR BECOMING

To the questions that we asked about education at the beginning of this chapter, science brings only very partial answers. The essential lesson that it teaches is the extreme richness of our potential and its plasticity. Intelligence is not something unchangeable; it is constantly developing.

Unfortunately, the language that we use tends, on the contrary, to lock each individual into a fixed definition. When we discuss a child, we are in the habit of enumerating his "gifts"; his various inclinations are interpreted as proof that he is "made for the arts," that he is "gifted at music," or that he has "a mathematical bent." This description is not necessarily absurd and may be confirmed by the course taken subsequently by the individual in question; to what extent do these words correspond to a reality, and to what reality?

It is clear that individuals are very different from each other. It is probable that their interest in and capacity for the various forms of intellectual activity are not identical. But what do we mean by saying that it is a question of being "gifted"? According to the dictionary, "to be gifted" is "to possess naturally"; in evoking a gift, we are assuming that this gift has been provided by nature. The only thing that nature gives us initially is a collection of genes; a particular trait is truly a "gift" if we can assume that it is a direct consequence of the activity of the genes with which we are endowed. Given what we said about the ontogenesis of the central nervous system and the determinism of certain intellectual defects, this statement can scarcely be proved except in the case of certain "negative" gifts, that is, certain defects.

It has not hitherto been possible to precisely link any specific intellectual capacity to genes. The most frequently mentioned case is that of music, the Bach family being sometimes cited as proof of the existence of a "gene for music," but it is impossible to disentangle a possible genetic effect from the very definite effect of the environment.

The existence of certain family patterns of specialization is not sufficient proof of a genetic origin. The impossibility of carrying out experiments on our species (there is not only the obvious ethical stumbling block; there is also the problem of time, since the generation span is so long) has, up to the present time, permitted the study of only a few, usually pathological, personality traits. We saw in the case of schizophrenia, which has been the most extensively studied in this field, how difficult it was to draw clear conclusions. And here we are talking about a psychic entity which, according to specialists, has a relatively objective definition. What difficulties would one not encounter in the study of traits as poorly defined as mathematical ability or a flare for music!

The specification and proof of a genetic mechanism require a great number of observations, carried out in the course of experiments conducted in accordance with rigorous protocols on clearly defined traits. None of these conditions can be satisfied in the case of intellectual activity; any talk about gifts is therefore never likely to get beyond opinions and ideologies. A scientist cannot but voice his doubts and react when his discipline is being improperly invoked in this regard.

The "gift" ideology, which locks each person into the often narrow set of capacities that nature is supposed to have given him, has been reinforced by the movement in favor of "highly gifted" children with exceptional ability, who are seen as victims of an excessively uniform educational system, and whose gifts, according to certain authors, should be better exploited for the benefit of all. That the excessive uniformity of teaching may be contrary to the extreme diversity of children and may be bad for some of them is not denied by anyone. But what do people mean exactly when they talk about the "highly gifted"?

We saw the difficulties associated with the term "gifted"; the addition of the word "highly" increases them. "Highly" has meaning only by contrast to "poorly"; it refers to a scale of measurement, and to a single scale. If one wishes to add "highly" to "gifted," one has to assume that the gifts in question can be measured globally by a number that sums them all up. This is the first condition for the word "highly" to have meaning; we are far from the commonly accepted view of intelligence.

The very definition of IQ guarantees us that only 0.4 percent of children score more than 140, 2.3 percent more than 130. All that is required therefore is that the bar be placed at the desired level in order to be able to say that "*x* percent of children are 'highly gifted'"; this, contrary to the way it is usually presented, is not a finding—it is a purely arbitrary definition. What we have said about the imprecision, about the instability, and, above all, about the uncertain meaning of IQ is therefore valid for the highly gifted.

But the real problem is not questioning the meaning of the term "highly gifted"; it is understanding why this theme has been so widely developed, with such ambiguous terminology.

For the zealots of these theories, it is a question of defending the idea that there is a natural hierarchy between people. The "highly gifted" are, in effect, only those individuals who are located at one extreme of an arbitrarily chosen scale of measurement, but they are presented as superior beings whose abilities are such that they constitute a valuable resource for society. Therefore society should, both in fairness to them and in the best interests of all, give them the best opportunities for perfecting themselves, the fastest careers, the positions with the most responsibility; they are to be the future leaders.

Such confusion of concepts leaves one bewildered: why would an above-average rate of intellectual development in childhood be a sign of global aptitude? In essence, it is a matter of precocity. There is no doubt but that, in extreme cases, this may lead to specific problems within the school system, but how could one possibly justify the linking of this precocity with prefabricated social success? The crucial error, implied by the very term "highly gifted," is the pronouncement of a global and definitive judgment on a child. It is too easy to cite cases of unprecocious children whose intellectual success was nonetheless brilliant: Beethoven, Darwin, Einstein, and Tolstoi were no doubt very lucky to have lived at a time when IQ had not yet been invented. The selective mechanisms that we have since developed would have streamed them toward the short circuits which would have quickly prepared them for the so-called active life: the first was "a hopeless case," the second was of "slightly less than average intelligence," Einstein's intelligence was "slow," Tolstoi was "neither hard working nor able."[14]

The fundamental mistake is in imagining that there is a preexisting gift, a talent given by the fairies, by God, or by genes. We are constantly developing. Let us listen, in conclusion, to J.-P. Sartre: "My madness protected me, from the very beginning, from being seduced by the 'elite,' never did I consider myself the lucky owner of a 'talent' . . . . [I am] a man made of all men, and whose worth is equal to, and equaled by, that of any one of them."

6.
Biology and Social Organization: Sociobiology

The organization of certain animal species, especially that of "the social insects," has always fascinated people. They discover structures seemingly very close to those they themselves have created to organize their own societies. Among wasps, bees, or ants, the behavior of each individual seems narrowly conditioned by rigorous social rules which define the roles of the various categories: workers, soldiers, queens, etc. The fairly recent development of ethology, the systematic study of animal habits, has shown this collective organization to be extremely widespread throughout the animal kingdom. Among certain fish and birds, the collectivity extends scarcely any farther than the procreating couple, within which the roles are allotted in such a way as to ensure protection of the young; these couples may last a season or, on the contrary, as with swans and geese, remain stable during an entire lifetime. Among many mammals, it is the group as a whole that seems to be organized in an often complex way as a function of the relationships between individuals, which we describe with terms such as domination, submission, altruism, aggressivity, and hierarchy, which have been defined in reference to human behavior.

However, no matter how exciting this is, the description of phenomena is not adequate for the development of a scientific discipline; it is necessary to try to propose some explanations: such is the ambition of a new branch of zoology, "sociobiology." Developed in the United States where it provoked violent reaction in university circles, it came to be widely known in France through popular magazines. A clear understanding of this new discipline, of its strengths and its limits, has important implications, because man is also an animal. Our ambition to understand the mechanisms to which other species are subject is strongly motivated by our desire to understand the mechanisms at work in our own species. An incorrect explanation for the behavior of a wild goose or a baboon is, of course, regrettable, but it can be easily corrected; it may be catastrophic if rules for our own behavior were to be based on this explanation. Let us therefore attempt to clearly specify the facts and arguments of sociobiology, eliminating the ideological or political contaminations that have impassioned, but also clouded, the debate.

AN OUTLINE OF SOCIOBIOLOGY

It deals with society, it deals with biology. According to the definition given by the creator of this discipline, Edward Wilson, in his lengthy book *Sociobiology: The New Synthesis* published in 1975, it involves "the systematic study of the biological basis of all social behavior." It is therefore an attempt to establish a link between the set of rules governing relationships between individuals within a group and the biological heritage of this group.

These rules can be described in a given epoch with great accuracy. But the aim of science is not only to describe, but also to explain them by imagining models dependent on the least possible number of parameters.

How and why were these rules gradually developed, adopted, and maintained? The study of reality cannot, in this instance, be complete unless we include an explanation of its dynamic; in order to proceed we must therefore analyze the process by which social rules are transmitted from generation to generation.

Wilson defines a society as "a group of individuals belonging to the same species and endowed with a cooperative organization." The reference to the "same species" means that these individuals are interfertile, that is, capable of producing individuals for the next generation. The process we are studying therefore encompasses that of reproduction.

Now in sexual species (and we are limiting ourselves here to this case), this reproduction constitutes a challenge to common sense, so much so that the meaning conveyed by the words used is the opposite to the reality that they are intended to designate. A unicellular organism capable of duplicating itself can "reproduce" or make a copy of itself. A sexual being has lost this power; all it can do is to enlist the help of a similar being to procreate a third one, which "reproduces" neither one parent, nor the other, nor even the average between the two; it constitutes a new entity.

The mechanism by which two individuals produce a third confronts our logic with a seemingly insoluble problem. The first attempts at explanation ascribed a major role to only one of the parents, usually the father, of course, who provided the "seed," the mother providing only the "soil," (this dysymmetry can still be found in some of today's elementary sex-education manuals for children). Quite belatedly, the theory according to which the child is prefabricated not in the sperm, but in the egg, was defined, but neither ovists nor spermatists could explain its obviously dual origin. As for the hypothesis proposed by Buffon and echoed by Darwin, according to which every elementary trait in the child is close to the average between parental traits, it implies the gradual homogenization of every population, which is contrary to observed fact.

This difficulty was not resolved until the nineteenth century. It took all of Mendel's naive audacity to claim that the "individual" is in reality "divisible". Each being received at conception a double collection of factors (we now say *genes*) which are active in him, while they themselves remain identical throughout his existence without changing each other. These inalterable and indivisible genes are transmitted to the children through a kind of lottery which, for each trait, designates either a gene of paternal origin or a gene of maternal origin. The concept "individual" is applied to a new object: it no longer concerns persons, but the genes that they carry.

All our thinking about transmission between generations must take account of this mechanism to which, unfortunately, our mind has difficulty in becoming accustomed: thirty-five years elapsed between Mendel's paper to a learned society in Brno in 1865 and the adoption, not without considerable reticence, of the concept of a gene by the scientific community.

Sociobiological discourse addresses three categories of objects: society, a group studied in its own right, characterized by its cooperative organization (in Wilson's own words); the people who constitute this group; the genes which these people carry. The evolution over time of these three categories of objects happens on three levels which it is useful to represent by a diagram, figure 6. We have represented two successive generations of the society being considered, S_t and S_{t+1}; an arrow links these within the "realm of societies," but this arrow has no biological meaning; the transmission of the biological material occurs within the "realm of persons," where each individual C of the S_{t+1} generation is a product of two individuals F and M of the S_t generation. However, here again, we are hardly capable of specifying the meaning of the arrows joining F and M to C; we can only correctly describe the mechanism involved by

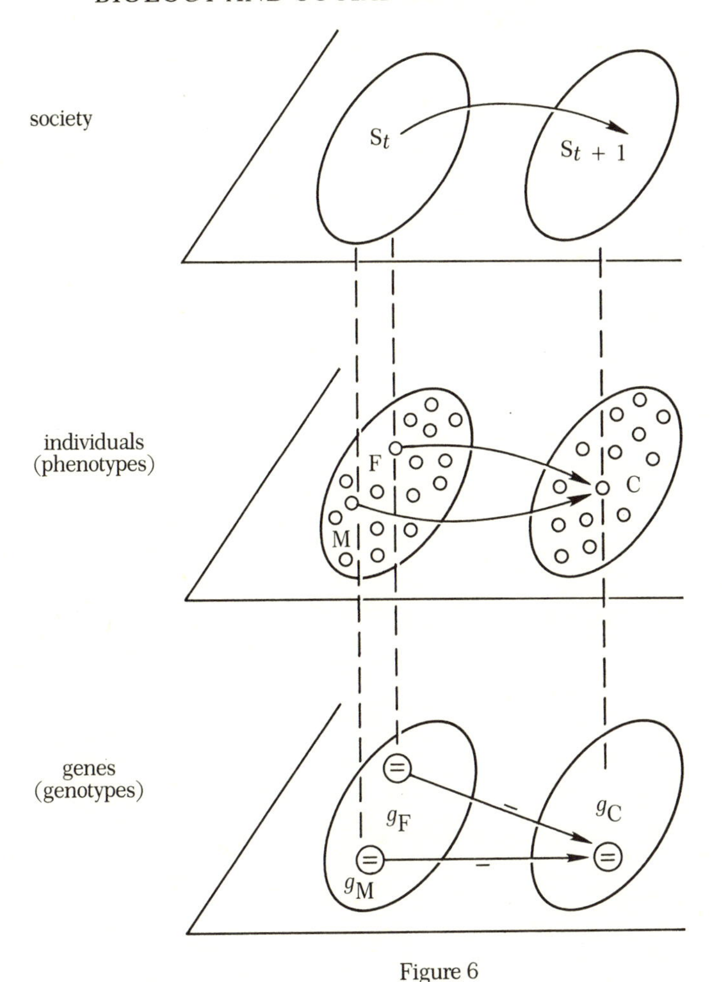

Figure 6

referring to the "realm of genes": we know that half of C's genes come from F and half from M, and that, from each of them, they were chosen at random.

However elementary this diagram, it seems necessary to an understanding of most of the debates triggered by the theses of sociobiology. It highlights the central difficulty: we know how to observe societies and individuals, but we know how to explain biological transmission only for genes.

Genetics has allowed us to specify the link that exists between a particular trait in a father and the same trait in his son by explaining how the trait in question is subject to a genetic determinism. In other words, it has replaced the arrow going directly from F to C with a more complex set of connections, but each stage of which is analyzable: $\{F - g_F - g_C - C\}$, where g_F is F's genetic complement. Let us consider the example of the disease called "cystic fibrosis"; the correspondence between genotypes and phenotypes has been established as follows:

Genotype	*Phenotype*
NN	healthy
Nc	healthy
cc	affected

where *N* and *c* designate the two modalities of the genes governing this trait. It is thus easy to explain how two healthy parents produce an affected child; all that is required is that both of them have the *Nc* genotype, an event with a probability that is easy to calculate.

However, such simple correspondences are relatively rare; they are found above all in the case of diseases linked to innate metabolic errors, such as the various blood systems. Usually, even for traits that are undoubtedly subject to a partially genetic determinism, the mechanisms are too complex, and involve too many interactions between the various genes or between the genes and the environment for it to be possible to propose a correspondence. We saw, for example, how difficult it is to specify a possible role of the genetic heritage in the case of the "trait" schizophrenia.

The aim of sociobiology is to replace, in similar fashion, the arrow going from S_t to S_{t+1} by a set of connections going through the collective genetic heritage. It therefore involves defining a genetic mechanism for individual traits with an influence on the structure and functioning of society—in other words, giving a meaning to the arrow linking g_F to F, then F to S.

This approach is perfectly legitimate: if it succeeds, it will make it possible to explain the gradual development of the observed social structure in conjunction with changes in the collective genetic heritage; now these changes are being studied by a discipline whose achievements are by no means negligible, population genetics. It therefore involves getting scientists with very diverse specialities to cooperate on a well-defined problem. The obstacles are considerable, as is shown by research on a type of behavior whose survival may seem paradoxical: altruism.

ALTRUISM AND SELECTIVE VALUE

"Altruistic" behavior, that is, putting the interests of the group or of another individual before one's own, cannot but lead to a lessening of the "selective value" of the latter. In the Darwinian sense of the term, this "value" corresponds to each person's capacity for transmitting the biological heritage (that is, genetic) of which he is the carrier. It is obvious that this transmission can only be hampered by inclinations toward self-sacrifice and dedication to others. Thus, altruism can be defined as behavior that lowers the selective value of one individual, to the benefit of the selective value of one or many other members of the group; similarly, egoism can be defined as the opposite attitude.

Now the study of many species, especially of social insects, shows numerous cases where such altruism seems to be found—for instance, the "soldier" castes among various species of termites or of fleas; these will even go as far as suicide to protect the group. If one assumes that this behavior is a phenotype governed, at least in part, by a genotype, the genes responsible should have been gradually eliminated.

This seeming paradox has given rise to much research; the diversity of explanations proposed demonstrates the difficulty of arriving at a "scientific" explanation in these areas.

A first approach was suggested as early as 1932 by the famous English geneticist Haldane: he pointed out that selective pressures affect not only individuals in competition with each other within a population, but also the various populations, both within the same species or from different species, who share the same ecological sphere. A population has, in itself, a selective value, independently of the selective values of the individuals who constitute it. The global value characterizes its degree of success in its struggle against the environment; for measurement purposes, researchers often use an undoubtedly arbitrary but objective parameter, which can, in general, be easily estimated—population size. A definition is then proposed: the selective value of a group is, in a given period, proportional to the rate of variation in its size during this period.

The two concepts of selective value thus defined, that of individuals and that of populations, are definitely not independent, but the link between them is complex. A group's advantage is generally due to its containing individuals with a higher value, but even more important to its survival is its capacity for structuring and for organizing itself to its own best advantage. When a population has achieved a high collective level of competiveness, thanks to an efficient structure that uses the complimentarity of individuals to the best advantage, any change in the balance between the individuals, as a result of a change in genetic frequencies, constitutes a danger. Competition at the individual level, which brings about an increase in the average selective values of individuals, can have negative effects on the selective value of the population within which it occurs and which is itself involved in an interpopulation competition within a particular ecological community: the success of the collective "project" requires that individual "plans" be forgotten.

In light of this, the future of a possible gene for altruism is the result of a multiplicity of interactions: the individuals that carry it risk being eliminated; its frequency in each population therefore decreases, but those groups where there is a lot of altruism are more competitive and triumph over the others. The simultaneous study of these two mechanisms must take account of the difference in their rhythms: the unit of time in competition between individuals is the generation, since nothing can manifest itself until children succeed parents; in competition between groups, this unit can be very small, since the elimination of one population by another can happen very rapidly.

The mathematical formulation of this process can be undertaken only by working with simplifying hypotheses that deprive the model of a good deal of its realism. Research carried out by means of computer simulations makes it possible to take better account of the complexity of reality, but so many parameters are introduced that it is difficult to draw clear conclusions.

Another research direction was outlined, mainly by W. D. Hamilton from 1964 onward, thanks to the introduction of the concept of kin selection. A very clear critical exposition of it appears in a book by Marshall Sahlins that was recently published in French.[1] When two individuals are related, some of the genes in each of them are copies of genes transmitted by their common ancestor. The existence of these "identical" genes, derived through successive duplications from generation to generation from the same ancestral gene, is the basis for the measure of relatedness (or kinship): the kinship coefficient φ of X and Y is the probability that two genes chosen at random at the same locus in both X and Y will be identical.

If X has altruistic behavior that brings about a reduction in his selective value, the resulting loss to the transmission of his own genes can be compensated for if this behavior increases the selective value of his relatives, some of whose genes are identical to his. Let *c* be the relative cost of this behavior for X, *b* the benefit to Y. Hamilton has shown that, in simple cases of relatedness, the frequency of the genes involved is increased, and not diminished by this altruism provided that

$$\frac{b}{c} > \frac{1}{2\varphi}.$$

In the case of two brothers or of a brother and his son, the coefficient of kinship φ is equal to 1/4. X's altruism is therefore beneficial on average for the genes that he carries if the advantage that it confers on X's brother or son is twice as great as its disadvantage to X himself.

This result can be illustrated by the classic story about children in distress on a boat. They are going to drown; should their father (or their brother) risk his life to save them? If he saves just one son (or one brother) at the expense of his own life, his genetic heritage loses out, since for each gene identical to his own that he saves in this way, he sacrifices two others; if he saves two sons, there is a balance; if he saves three, there is a gain. Our genes must therefore, in the interests of their survival, lead us to egoism in the first case, to indifference in the second, to (seeming) self-sacrifice in the third.

Looked at this way, the future of a gene is governed not by the individual selective value of the person who carries it, but by the sum of this value and of all the effects that the behavior induced in X by this gene has on the selective value of the various relatives—what Hamilton calls "inclusive fitness."

Other mechanisms have been imagined; it can nevertheless be noted that all these arguments rely on the initial models of neo-Darwinism, based on the definition of a "selective value for the genes." Now most of these models have had either to be abandoned or to be made more complex. After a period of triumph when neo-Darwinism had managed to reconcile the mechanism of natural selection imagined by Darwin and the mechanism of transmission from parents to offspring discovered by Mendel, this discipline encountered a major obstacle in the sixties: it proved to be incapable of explaining the high level of genetic polymorphism in populations.

Apart from exceptional cases, natural selection has a "purifying" effect; eliminating bad genes, it leads to a gradual homogenization; little by little, all individuals come closer to the ideal type corresponding to their environmental conditions. However, the study of the genetic structure (especially with the help of electrophoretic techniques) shows that reality is the opposite to this prediction: for a third or half of all traits, natural populations are not at all homogeneous—they are "polymorphic."

This contradiction between the data gathered from actual populations and the consequences of the theoretical models that had been developed is the main challenge facing researchers in this domain.

In the next chapter, we will see how the approaches being explored with a view to resolving this paradox are leading to quite a radical revision of neo-Darwinism. In this context, it is clear that the arguments proposed by the pioneers of sociobiology are in danger of having to be profoundly reshuffled. This, to tell the truth, is a prospect to which scientists look forward with pleasure.

Like any discipline in its early stages, sociobiology is in search of precise concepts and of coherent explanatory models. The case of possible "genes for altruism" shows how far it is, in its present phase, from a sound well-founded theory.

Valéry said of history that if "offers examples of everything," which makes it possible to back up any statement with a case from history. The same is true of the social structures developed by the various species; the attempt to explain everything leads to explaining nothing. The sociobiologist is aware of this; his attempts at shedding a little light on a terrifyingly complex reality can only be encouraged, at least as long as the rules of logic are rigorously adhered to.

This rigor is currently seriously jeopardized by the excesses of the media which present as definitive theories provisional explanatory models that have been developed on the way, for want of better. Sociobiology's success is such that the risk of deformation is considerable. It would not matter very much if it were merely a question of discoursing on the genetic determinism of the organization of the termite or on the biological foundations of the hierarchical structure of *Papio anubis*. But treatises on animals really excite us only in so far as we make inferences from them about man. References to sociobiology are usually aimed at forming an opinion about the organization of our societies.

What becomes of sociobiology when its subject is no longer some animal, far removed from us like the ant, or close like the chimpanzee, but man himself?

THE SOCIOBIOLOGY OF MAN

Wilson's work, so rich in facts, in reflections, and in references, contains twenty-seven chapters; the first twenty-six deal with theoretical models and animals, the twenty-seventh with man. At the outset, the author warns that there is a change of subject, because "*Homo sapiens is ecologically a very peculiar species*." Then he tries to apply to man certain theories developed with animals in mind. It is a fascinating exercise and may lead to some very useful reflections; it thus makes it possible to uncover those aspects of our behavior that are no doubt only a distant residue of an attitude that became entrenched in the course of the evolution of the species from which we are descended, whereas we could naively see them as a purely human creation.

Man is, to be sure, an animal; like all animals, he develops according to the program encoded in his genes; it is perfectly legitimate and may eventually be productive to study the links between his genetic heritage and all his activities, including the most elaborate ones. However, one has to avoid taking as proof what is only pure analogy.

In order to understand what is specific to human sociobiology, it is convenient to return to our initial diagram. We saw that biological transmission can be understood only at the genotypic level: the arrows uniting g_F or g_M to g_C have a precise meaning; those uniting F or M to C cannot be clearly defined.

That is true of biological transmission, and therefore it is equally true of the transmission of behavior, in so far as animals are "genetic automats," in the words of A. Langaney.[2] For many species, this automatism has gradually been attenuated, and part of their behavior is acquired through contact with other members of the group: in certain birds singing is strictly innate; in others learning is necessary. However, the human species is a special case in so far as it attaches enormous importance to learning and has moved far away from its status of genetic automat. Of all newborns, the human newborn is no doubt the most helpless, but also the most capable of learning: "The specificity of the human newborn is in being a totipotent nonentity," according to Langaney. This potential has been increased by humans through language, through writing. For man, infinitely more than for any other animal, an arrow can be drawn and defined between F or M and C, but it is no longer a biological transmission that is evoked—it is cultural transmission (see figure 7).

Sociobiology's main focus is therefore quite different depending on whether it is considering animals (among which a major part of the transmission determining changes to society occurs at the genotypic level) or, on the contrary, humans (this transmission is mostly cultural and happens at the "level of individuals").

In these circumstances, anything that will have been proved in one case can only be transposed to the other as an image, not as proof.

Unfortunately, some self-interested users of sociobiology are looking to it for "scientific" arguments to defend opinions that result entirely from their personal choices. Michel Rouzé takes the following from what he calls "sociobiological literature": "Capitalism, like competition and self-interest, is encoded in our genes; similarly, hostility toward immigrants or the domination of woman by man are encoded in them; altruism cannot be extended to all humankind."[3]

One could continue almost forever with this nonsense, which consists of peremptory but gratuitous state-

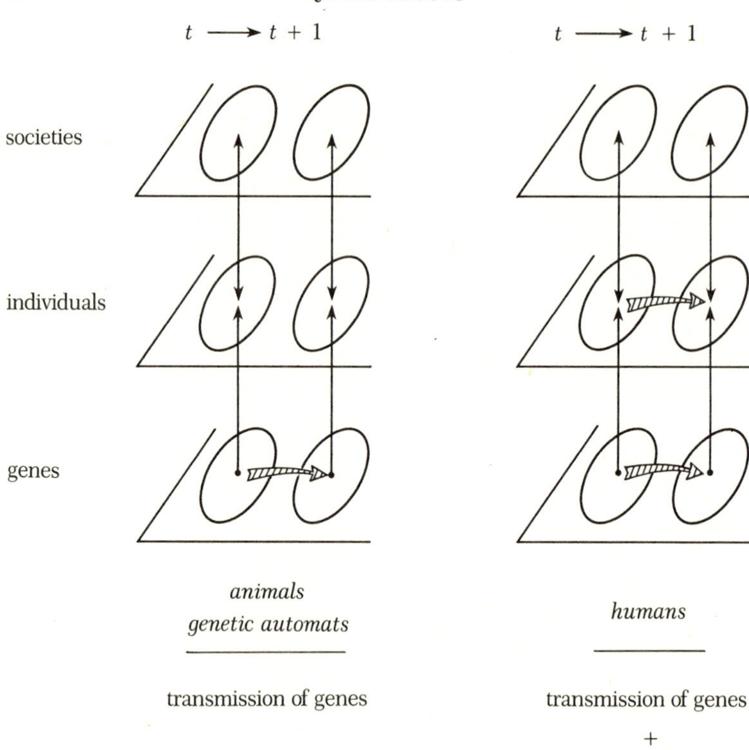

Figure 7

ments based on analogies, and these analogies can only be proposed by denying human specificity. Sociobiology, a perfectly legitimate discipline when it defines its scope and subject and attempts to better understand how certain social structures were instituted among animals, becomes, in the hands of some, a tool of oppression. This kind of deflection of a scientific theory is not in any way exceptional: in the nineteenth century, Darwinism was quickly used to justify the exploitation of certain social categories or certain peoples. According to "social Darwinism," this exploitation was only an instance of the struggle for survival, which undoubtedly had highly unpleasant consequences for some, but which was necessary for the survival and progress of the group. Darwin himself, swayed by the mentality of his times, advocated "eliminating those laws and customs that prevented the most capable from succeeding" and, consequently, the least capable from being eliminated.

The aberrations to which genetics has led have been far more monstrous. Let us remember: Professor von Verschuer, director of the Berlin Genetics Institute, congratulated himself in 1941 that "the Head of the German Ethnoempire is the first statesman to have made of hereditary biology a guiding principle for state affairs." And this same scientist, this geneticist, stated: "The present policy requires a new and total solution to the Jewish problem," and he advocated a similar solution to the tzigane problem.

Remember that in France, even a doctor, a member of the medical faculty at the University of Paris, proposed to "improve the biochemical index of the French race" by revoking the French nationality of individuals in the B blood group and by submitting the AB to a psychotechnical examination.

To be sure, nothing of this sort is being proposed today. However, we see attempts at perverse uses of science which must be halted before they become harmful: the grotesque proposal to improve the species with sperm from Nobel prizewinners and the meaningless statements about the heritability of intelligence or the genetic determinism of altruism are preparing the way for a renewal of eugenics.

The genetic origin of certain structures in human societies or of certain feelings expressed by people in society (aggressivity, altruism, the desire for ownership and for power) is no more than a hypothesis based on vague analogies with animal societies. In reality, we are unable to look at these societies in a completely unbiased fashion because we are influenced by what we know of our own organization: the fact that we call various categories of bees "queens," "workers," or "soldiers" clearly shows that we have projected onto them our own ideas about group structure. It is not surprising that in return we notice that our societies show analogies with those of insects! In the face of attempts at extrapolation aimed at presenting, for instance, hierarchical structure as a necessity linked to our genetic heritage, it is important to react by stressing the specificity and lack of content of human sociobiology.

The role of science is to give man new means of action; even more than that, it is to help him to give direction to this action. Lucidity is more necessary that efficacity. However, this lucidity has scarcely any weight if it remains confined to laboratories.

Science can be a tool for liberation, but it can also become a tool for oppression; in order to avoid this deflection, those responsible for its progress must share their knowledge by expressing themselves in such a way as to be understood by all.

7. The Evolution of the Living World: Facts and Explanatory Models

Living beings, whose infinite variety amazes us, are all related.

From the primitive forms that appeared in the oceans a little over three billion years ago (our planet earth was already more than a billion and a half years old), innumerable beings proliferated. The life span of each of them was in general very short, but they were capable of a certain "survival" by producing descendants in their own image. This capacity for reproduction made the maintenance of life possible in spite of the evanescence of individuals. Achieved originally by duplication, this reproduction was, by chance, sometimes slightly different from the original. Innovations appeared, creating new types, endowed with previously unknown possibilities. The most decisive of these innovations was no doubt the appearance (one or two billion years ago) of a very peculiar mode of reproduction: the duplication of an autonomous individual was replaced by collaboration between two individuals associated to make a third, which, among other consequences, speeded up diversification. New organs made it possible to escape the initial marine environment, to conquer first the lands that had emerged and then the skies. Heterogeneous groups became differentiated, and the tree of the living world became gradually ramified, leading (provisionally) to the current peopling of our planet earth by an incalculable number of living beings divided into several million species, of which ours is one.

Presenting the past in this kind of *evolutionary* framework is no longer just one theory among others; it can now be presented as fact, so numerous are the findings that support it:

—Similarities between anatomical forms: the basic resemblance between the organs of the various species, which becomes all the more striking as the analysis is more precise, led people to suspect a link between these species. As early as 1721, Montesquieu proposed this idea which, after being adopted by Maupertuis and Buffon, was developed and generalized by Lamarck; at the beginning of the nineteenth century, Lamarck developed an explanatory system from it—transformationism.

—Similarities in the development of embryos: during the early phases of its development, a human embryo, for instance, bears a very close resemblance to that of a cat, a tortoise, or a fish. It even develops the rudiments of organs which, if they were to continue to develop, would be useless to it, because they do not correspond to the conditions under which its life will be lived; such is the case with the pharyngeal openings that are necessary to bronchial breathing in fish, but that have no function in humans.

—Paleontological findings: for certain lines, the fossils that have been found make it possible to describe in continuous fashion the gradual transformation of species over a long period of time.

—Physiological resemblances: the structure of cells is similar in all species; these cells accomplish the transfer and storage of energy using the same substances. Advances in biochemistry have shown that these similarities are also found at the molecular level—the structure of polypeptides (cytochrome C, hemoglobin, fibrinopeptides) is very stable from one species to another.

—Finally and above all, the unicity of the genetic code: the correspondence between the bases of the DNA and the amino acids is the same for all those species within which this correspondence could be established. The language of chromosomes is one throughout the entire living world.

Science, in dealing with this fact of evolution,[1] must not only describe it by retracing as best it can the history of living things and by detailing its end—it must also explain it. Behind the chronicle which we are gradually writing in increasing detail, we must imagine the processes of which it is the result.

Matters were clear with a static hypothesis: everything was immutable, and the only happening was the creation, simultaneous or otherwise, of all species by divine power. However, when we accept that things change and become transformed, we have to understand how these changes come about.

REPRODUCTION AND PROCREATION

The essential mechanism of the evolutionary process is obviously the transmission of the biological heritage from one generation to the next. This mechanism is simple in the case of unicellular organisms capable of "reproducing" themselves: changes can come only from errors occurring during this copying process, *mutations*.

But in organisms with biparental reproduction this mechanism, as we have already stressed, defies common sense. Confronted with this problem, some cultures (especially the most "advanced") have resolved it by denying it, by imagining that the child was prefabricated in the gametes of one of the parents, the egg or the spermatozoon; only one of the parents is thus supposed to play an active role. Other cultures, taking account of the symmetry of the parental roles, proposed various hypotheses (in particular, that of pangenesis, by Darwin); they assume in general that, because of fusion of the parental contributions, each trait in the child is, with one or two random exceptions, the average of the parental traits. However, the consequence of a theory of this kind is an ineluctable uniformization of all traits within every population, which is contradicted by reality.

The way out of this paradox is the hypothesis, proposed as early as 1865 by Mendel, rediscovered in 1900, and definitively accepted since then, concerning the transmission of inalterable elementary factors in the individual. Remember once again that, in light of this hypothesis, there is no such thing as the transmission of traits, but only the transmission of the genes (half from each parent) that determine them; this is something totally different and renders all previous attempts at explaining evolution obsolete. Procreation is no longer reproduction; it is the assembling of a double collection of genes taken half from the father and half from the mother.

From one generation to the next, changes come from two sources:

—one, systematic, is the recombination of the genes received from the two parents, a recombination that makes each being unique;
—the other, random, is the copying error that gives rise to a new gene with a new type of activity, that is, a new *allele*, by mutation.

PHENOTYPE AND GENOTYPE

The very formulation of the problem of evolution is thus totally changed. The entities where evolution can be detected remain the same: a particular form of an organ, a particular function, a particular metabolism. However, the entity that is transmitted is a gene affecting this form, this function, or this metabolism.

It is impossible to overemphasize the difficulty introduced by the necessity for this double discourse—one focusing on what we can actually observe at the phenotypic level, the other on mechanisms that we can explain at the genotypic level. The link between these two levels has been clarified only in those few special cases (blood group or enzyme systems, metabolic disorders) where it has been possible to establish a strict genotype-phenotype correspondence; usually, this link can be explained only by means of very ambiguous statistical parameters, such as those linked to the concept of heritability, which harbors so many pitfalls. For about half a century, innumerable works have attempted to take account simultaneously of the observations (the phenotypes) and of the mechanisms (the genotypes). The combination of these attempts is usually presented under the title of "synthetic theory of evolution." This rather pompous title risks giving the false impression that an imposing monument, lacking only the finishing touches, has been erected. In reality, we find ourselves in a building site cluttered with scattered materials where the workers are remarkably active, but where the various contractors are engaged in noisy quarrels.

As often happens, this new discipline has developed, not according to its own inner logic, but according to the directions set by previous disciplines. The revolution brought about by the introduction of a new concept, that of a gene, and by the rejection of the old one, that of the transmission of traits, was experienced first of all as an attack, an unexpected difficulty to be circumscribed and resolved. Considerable effort was expended with a view to saving, with the least possible amount of adaptation, the old theories. In this difficult enterprise, Darwinism weighed heavily; it is in reference to it that the various directions taken by the new research were situated. Some of these are presented as neo-Darwinian, others as non-Darwinian.

It now seems possible to free ourselves from the tangle of confrontations beetween theoreticians, to free ourselves also from their sequel and to attempt an exposition, which, in going logically from simple to complex, actually goes contrary to the history of the development of ideas; we must begin with the most recent in order to logically introduce the various concepts involved.

POPULATION GENETICS

The successive generations of a population are composed of individuals, all of whom are distinct, unique, and stable. From parent to child, no "evolution" occurs since there is creation of an entirely new being because of genetic recombinations. "What" evolves is therefore neither an individual nor a collection of individuals, but a collection of genes. The definition of this collection and the study of the factors which change it from one generation to the next are the subjects of a rather peculiar branch of genetics: *population genetics*.

Rather peculiar, since many of the researchers involved in it know nothing of the white coats, test tubes, and microscopes of laboratories; they are interested less in the genes themselves than in their overall structure, that is, their frequencies. These frequencies gradually change through a process where "chance" is an essential factor: the chance appearance of a mutation, the chance formation of couples, and above all, the chance in the choice of paternal or maternal genes during the making of the gametes.

We saw in chapter 2 how terribly encumbered with various meanings this word "chance" is. Without repeating the discussion about these meanings and their implications for our vision of the determinisms at work in the universe, let us say that the mechanisms that we are studying are such that their description requires the use of probabilistic arguments: the subjects of discourse are parameters (frequencies, number of alleles) for which we can study only the laws of probability.

Finally, the population geneticist is a manufacturer of mathematical models, whose validity he studies either by carrying out pseudoexperiments using simulations (of the Monte-Carlo variety usually, to take account of aleatory phenomena: this is a game where modern computers are remarkable partners; see box 3), or by comparing his results with the observations of biologists. In this last stage, difficulties arise, since it involves confronting the two "levels of discourse," that of phenotypes and that of genotypes.

Box 3

Many processes can be described only by assuming that chance plays a part in determining the state of the system at a given time t, taking account of its state at a previous time $t - 1$.

Suppose that the "system" under study is a population composed, in a certain generation g, of 50 people with, for the ABO blood group system, 100 genes distributed as follows: 25 *A*, 10 *B*, 65 *O*. The making of generation $g + 1$ consists of a series of lotteries: for each individual chance designates two genes, each of which has a probability of 0.25 of being *A*, 0.10 of being *B*, 0.65 of being *O*. Any result is possible (for instance, with a probability of $0.65^{100} \approx 2/10^{19}$, the entire $g + 1$ generation will be homozygous *OO*, which is highly improbable but not strictly impossible). To simulate this process, the computer assigns to each individual two numbers between 0 and 99, determined according to a pseudorandom mechanism which gives to each value the same probability of being chosen.

It assigns the *A* gene if the number is between 0 and 24, the *B* gene if it is between 25 and 34, the *O* gene if it is greater than or equal to 35.

In this way, one can gradually simulate the chance transformation of the genetic structure of a population over a long period, with the possibility of considerably reducing the duration of the observation: in a few minutes, a computer carries out a pseudoexperiment that supplies as much information as the observation of a population over a few hundred generations.

This obstacle disappears when the trait under study is so directly linked to the genes that govern it that the phenotype-genotype correspondence is clearly understood. Such is the case when this trait is the structure of one of the chains of amino acids that constitute the essential materials of living organisms. Just as the genetic information is supplied by the molecules of DNA in the chromosomes, which are long chains of which each link is one of the four bases, A, T, G, or C, so too the principal functions of our organism are maintained by polypeptides, chains of which each link is an amino acid, chosen from among twenty different types.

These amino acids are the twenty letters of an alphabet with which the polypeptides are "written," just as the nucleotide bases are the four letters of the alphabet with which the DNA is written; the genetic code makes it possible to make one amino acid correspond unambiguously to each "triplet" of DNA bases (on the contrary, to one amino acid may correspond several triplets). The study of evolution should therefore focus primarily on all these structures; unfortunately, our knowledge of them is relatively recent, and it has been scarcely ten years since advances in biochemistry have made it possible to gather enough information to enable a specific discipline, the "molecular genetics of populations," to develop.

1. The Evolution of Molecular Structures

The study of evolution in this discipline comes down to analyzing the transformation of a given polypeptide chain in the course of the differentiation of species. To do this, polypeptides common to many species must be compared with each other. Such is the case with the various chains of hemoglobin that are found in almost all vertebrates, with cytochrome C, present in almost all aerobic organisms, with histones, the principal components (with DNA) of the chromosomes, or with fibrinopeptides, residues of blood coagulation.

The comparison of sequences of the same chain in different species demonstrates that they have similar structures and differ by amino acid substitutions, all the more numerous as the divergence of the species being compared is further in the past. Thus, for the hemoglobin chain which in man contains 141 amino acids, only 18 of them are different in the man-horse comparison, 16 for man-mouse, 62 for man-trition, and 68 for man-carp. The evolution of this "trait" therefore involved an accumulation of mutations each corresponding to the substitution of one amino acid for another.

THE RATE OF MOLECULAR EVOLUTION

The essential finding is that, for each peptide, the rate of these substitutions is noticeably constant regardless of the pair of species being compared, but this rate is very variable from one peptide to another.

A simple procedure (see box 4) allows us to calculate the rate of change, λ, per site and per year. From the information on sequences provided by biochemists, we know the proportion k of amino acids identical for two species and, from paleontological data, the length of time T that has elapsed since their divergence. One finds that the parameter λ is practically constant for a given polypeptide. According to Masatoshi Nei, one obtains:

	Rate of substitution per million years and per 1,000 amino acids
Fibrinopeptide	9.0
Hemoglobin chain	1.4
Cytochrome C	0.3
Histone IV	0.006

In the course of the millions of years during which the differentiation of species was taking place, each polypeptide chain was gradually changed at a rate that seems, on average, constant: whether in aquatic or aerial species, fish or mammals, the hemoglobin chain α changed at a constant rate—likewise cytochrome C, but the speed of transformation of the first is four or five times greater than that of the second.

How are these findings to be explained? The simplest hypothesis is the following: spontaneous mutations oc-

Box 4

A polypeptide chain is constituted by a sequence of n amino acids (in the case of the chain for human hemoglobin, $n = 141$). As a consequence of mutations affecting the DNA, the amino acid occupying each of the n sites may be replaced by another. Let λ be the probability that this event will happen, at a given site, in the course of one year; the probability that it will not happen in the course of x years is $(1 - \lambda)^x$.

If two species have diverged for T years, the probability that the amino acids at the site under consideration will be identical is therefore $(1 - \lambda)^{2T}$, if one discounts the possibility of two successive mutations recreating the initial state. If, for a given polypeptide, the percentage of identical amino acids between the two species is k, one can therefore write:

$$(1 - \lambda)^{2T} = k;$$

hence, since $\lambda < 1$,

$$\lambda \cong -\frac{1}{2T} \log (k).$$

For the α chains in man and carp, paleontological data allow us to estimate T at $350 \cdot 10^6$ years, and observation of sequences of amino acids gives $k = 0.49$; hence $\lambda = 9.5 \cdot 10^{-10}$.

For the α chains in man and horse, $T = 70.10^6$ years and $k = 0.87$; hence $\lambda = 9.7 \cdot 10^{-10}$.

For the α and β chains in man, $T = 500.10^6$ years, and $k = 0.41$; hence $\lambda = 8.9 \cdot 10^{-10}$.

cur at a constant and uniform rate μ per amino acid site and per year, or $n\mu$ for an entire polypeptide chain comprised of n amino acids. Among these mutations, there are some that change the structure of the chain in such a way that it can no longer play its normal role; individuals undergoing these mutations have diminished chances of survival and of procreation; the mutated gene is eliminated sooner or later, and the probability that it will spread throughout the population is so slight that it can be considered nonexistent; the other genes, produced in this manner by mutation with a frequency of $\alpha n \mu$ (where α is the proportion of unfavorable mutations), are considered *neutral,* that is, the chances of spreading throughout the population are equal for all of them.

With this hypothesis, proposed for simplicity, the only mechanism for the transformation of the genetic structure, apart from mutations, is *random drift:* because of the very mechanism of genetic transmission, the frequency p of an allele in a generation is a random variable, that is, a variable whose value can be predicted only in the form of a law of probability. It can easily be shown that the probability of this variable is equal to the frequency P_0 of this allele in the preceding generation and that its variance is inversely proportional to the "genetic size" of the population (in general, close to the total number of individuals of procreating age).

These random fluctuations can, in the long run, have only two outcomes: either the new allele disappears, or else it replaces all the others, and it becomes fixed. The differences that we find between the polypeptides of the different species are the result of such fixations.

When a new neutral allele appears in a population consisting of N individuals, it is present in one copy among a total of $2N$ genes; all the alleles present being assumed to be neutral, the probability that one of them, and not another, will eventually become fixed is equal to $1/(2N)$. Each year the number of new alleles that appear is equal to $2N \times \alpha n \mu$; the number of those that will become fixed, which we have previously represented by λ, is therefore

$$\lambda = \frac{1}{2N} \times 2N\alpha n\mu = \alpha n\mu.$$

The rate of evolution is therefore independent of population size.

This very simple result, published in 1968 by Japanese geneticist Motoo Kimura, provides a perfect explanation for the constancy of the rate λ for each protein. It also explains the variations of λ according to the proteins; the proportion α of not unfavorable mutations is greater or lesser depending on the functional rigidity of the mechanism in which the protein is involved; for histones, which play a crucial role in the binding of the DNA, λ is low; on the contrary, for fibrinopeptides, whose function seems insignificant, λ is very high.

This comparative study can be pursued further by calculating the rate of fixation λ, not for a polypeptide chain as a whole, but for certain fragments of it. Again, one finds a relative constancy for each fragment regardless of the species being compared, but great variability depending on whether sensitive or nonsensitive zones of the chain are involved; according to Kimura and Ohta, the zone of hemoglobin chains linked to the heme evolved ten times less rapidly than the surface zones.

It is possible to pursue the comparison of rates of evolution still further by considering not just the mutation of amino acids, but that of the very bases of DNA. It is known that the genetic code gives the principal role to the first two letters of each triplet, a change of the third often leading (in about 70 percent of cases) to a synonymy. One can therefore expect a higher frequency of mutations affecting the DNA at this position, since these mutations are usually silent, therefore with no effect on the functioning of the organism. This is, in fact, what is observed.

THE MAINTENANCE OF MOLECULAR POLYMORPHISM

In an isolated population subject both to mutations and to drift, these two factors have opposite effects: the former increases polymorphism (that is, the genetic richness of the group) by constantly supplying new alleles, while the latter decreases it by making some of them disappear at random.

The results of these two factors can easily be studied by means of recurrence equations describing the evolution of the identity of genes from one generation to the next. One of the best measures of the intensity of polymorphism is H "heterozygosity" (this awkward term denotes either the proportion of individuals with two different alleles at one given locus,[3] or the proportion, in a given individual, of heterozygous loci). A simple argument makes it possible to establish that, at equilibrium, this heterozygosity is such that

$$H = \frac{4Nv}{1 + 4Nv}$$

where v is the rate of appearance of neutral alleles at each generation.

This time the size of N of the population intervenes, but not as a separate variable; the decisive parameter is the product Nv. If one assumes that $v = 10^{-6}$, which seems a typical value, the expected heterozygosity is only 4 percent for a population size of $N = 10^4$, but reaches 29 percent when $N = 10^5$, and tends toward 1 when N is greater than 10^6.

Direct observation of actual heterozygosity (two proteins being different as soon as a site is occupied by two distinct amino acids) would require complete anlaysis of the sequences of the protein under study in a large number of individuals, which would be extremely costly. Usually scientists settle for a much easier type of observation using the electrophoretic technique, but the latter detects only about a quarter of amino acid substitutions.

Extensive literature is devoted to the description of the electrophoretic polymorphism that has been observed in different species. An excellent synthesis of the information supplied by 180 publications was published by Eviatar Nevo.[4] This information concerns 243 species belonging to 37 orders; in each species, from 12 to 71 loci were studied. The average heterozygosity per group of species and its typical variation between species are given in table 3. H, whose average value is 7.4 percent, varies from 3.6 percent in mammals

Table 3
Heterozygosity of the Various Species

Number of species observed	*H*	*Average as a percentage*	*Standard deviation as a percentage*
Plants	15	7.1	7.1
	Invertebrates		
Drosophila	43	14.0	5.3
Other insects	23	7.4	8.1
Others	27	10.0	7.4
Total	93	11.2	7.2
Vertebrates			
Fish	51	5.1	3.4
Amphibians	13	7.9	4.2
Reptiles	17	4.7	2.3
Birds	7	4.7	3.6
Mammals	46	3.6	2.5
including man	—	6.7	—
Total	135	4.9	3.7
Overall total	243	7.4	5.1

Source: According to E. Nevo.

to 14 percent in *Drosophila*, which corresponds, using the equation between H, N and v, to sizes of N varying between 50,000 and 160,000. These orders of magnitude are acceptable, especially if one takes account of periods where, because of environmental changes, species sometimes suffer "catastrophic" reductions in their numbers.

The differences found in the levels of heterozygosity of the various species can only rarely be explained by the environments in which they live: to be sure, the amphibians, alternating life on land with life at sea, have a higher H (7.9 percent) than that of land vertebrates (4.1 percent) and aquatic vertebrates (5.6 percent in fresh water, 6.1 percent in salt water), but those fish that alternate between salt and fresh water have a very low H (2.5 percent). For invertebrates, H is not significantly different on land (10.1 percent on average) and in the sea (12.4 percent). One might, on the contrary, have predicted that species living in fluctuating environments would have a greater reserve of genetic variability and therefore much higher heterozygosity.

By contrast, vertebrate polymorphism is markedly higher on the mainlands (H = 8.7 percent) than in the islands (4.7 percent), which is explained by the influence of the size of N (the difference is in the same direction, but not significant for the invertebrates: 14.3 percent and 13.9 percent). Groupings by zone show that tropical species are, on average, more polymorphic (H = 10.9 percent) than species spread over several zones (H = 9.4 percent), themselves more polymorphic than the temperate species (6.6 percent), but the differences between species within the same zone are much greater than the average differences between zones. Taken as a whole, these findings do not therefore make it possible to refute the "neutral" hypothesis.

PHYLOGENETIC TREES

Knowledge of the current sequences of a given protein in different species makes it possible, by granting a certain arbitrariness in the choice of argument being used, to reconstitute a possible phylogenetic tree. The most probable tree is one that minimalizes the number of mutations between the various "knots," representing either the points of divergence between species or their current state. The few proteins currently available lead to trees that are very similar and above all very close to those that had been based on the study of fossils. This convergence between two disciplines whose concepts and techniques are very different from each other is especially remarkable.

For many groups of species, the paleontological data were very inadequate. Moreover, comparisons between the different phyla or the different realms were impossible: the classical phylogenetic trees were therefore necessarily fragmentary. On the contrary, molecular structures permit comparisons between the most distant species: the reconstitution of the tree of the whole of the living world can, thanks to them, be attempted.

"NON-DARWINISM"

In the description of this theory of molecular evolution, the word "selection" has not been uttered up to now, which justifies its being qualified as "non-Darwinian." This somewhat provocative qualifier has often been equated with "anti-Darwinian" and has stirred up much lively debate. In fact, the concept of selection is not entirely absent, since it was necessary to assume at the outset that certain genes that appeared by mutation are "unfavorable" and are quickly eliminated; the way this elimination actually happens is, to be sure, natural selection, which affects individuals endowed with these genes.

However, we are a long way from the selection described by Darwin and which was one of the essential factors taken into account by the theory developed from 1920 onward under the label "neo-Darwinism." This attempted to classify individuals on a scale of values: "selective value." The neutralist theory relies only on negative selection, capable of eliminating; the noneliminated genes are assumed to be all equivalent and these are, apart from exceptions, the only ones that we encounter in the real world. The term "neutralist theory" is no doubt inappropriate if we are referring to all the genes that have appeared as a result of mutation; it is appropriate if we are referring only to those genes that are present in the living world around us, differentiating species from each other, just like individuals within species.

This supposed neutrality of the various alleles must be interpreted less as a statement about a property of these alleles than as the result of a preterition. The goal of a theory is to take account of observed reality by introducing the minimum number of explanatory variables. It is possible, at least as a first conjecture, to take account of the evolution of molecular structures by having recourse to only two parameters: the rate of unfavorable mutations and the size of the population. Why introduce a concept as delicate as that of selective value, when it is not necessary?

The neutralist does not claim that two polypeptide chains have strictly the same effect on individuals' capacity for survival; he notes that the hypothesis of the absence of a significant difference makes it possible to explain what he observes. It is not a question of demonstrating that alleles are neutral, but of eventually finding a flaw in this hypothesis by proving that they are not. The burden of proof, a jurist would say, is in the opposite direction.

If Darwin's eye had been able to distinguish not the shape of finches' beaks in the Galapagos, but the sequences of their hemoglobin chains (which is, after all, an equally interesting trait), his thoughts on the mechanisms of evolution could have taken an entirely different direction. A century and a half later, our eyes, with the help of various instruments, look at reality in new ways. It is normal that our ways of thinking should also be new; the theory of molecular evolution is non-Darwinian in so far as it can do without the concepts introduced by Darwin.

However, the polypeptide chains are not the only ones to have evolved. We will see that matters are less simple when we study "traits" whose link with the genetic heritage is less obvious.

2. *The Evolution of Phenotypic Traits*

AN ALTOGETHER DIFFERENT PROBLEM

Those traits that strike our senses directly, those that led to the first speculations about the relatedness of the various species and their common origin, color, shape, and size, are usually linked in a complex way with the genetic heritage. They are, on the other hand, often linked in an obvious way to the capacity for survival and procreation. Stags with powerful horns outdo their rivals during the sharing of females, while those whose horns are over developed risk being hampered in their movements; to each variation of this trait one can thus associate a parameter measuring the value of the individual in the *struggle for survival* ("survival," in this expression, meaning more the transmission than the maintenance of life); this parameter is the "selective value."

This is the starting point of Darwin's thesis as presented in the *Origin of Species:* of all those who are born, some are supernumerary, given the limits to our resources; those that "*have some advantage over the others have a better chance of surviving and of procreating their own type.*" Unfortunately, if the beginning of this sentence is obvious, and even tautological, the end is completely wrong: in sexual species, no individual "*procreates his own type*"; each one can only contribute to his child's "type" by supplying half of the genes that define his own type.

This is not a mere nuance; we are confronted here with two conflicting visions of biological transmission. To understand the evolution of the trait "length of horn in stags," it is not enough to assign a selective value to each expression of this trait; we must also understand how it is transmitted. It is therefore necessary to establish a link between the genes, which are transmittable, and the trait expressed, which is not transmittable. The aim of neo-Darwinism was to reconcile these two research directions by synthesizing Darwin's argument, which deals exclusively with phenotypes, with the mechanism for transmission of the biological heritage, which deals only with genotypes. In many cases, this goal was achieved, but sometimes the hypothesis used suffered from a lack of realism that can justifiably be denounced.

The difference between the evolutionary mechanism at work on phenotypic traits and that at work in molecular structures has not always been sufficiently stressed. Consider the following contrast: we saw how the rate of transformation of these structures is uniform from one species to another; on the contrary, the evolution of shapes occurred at widely varying rates. Man's cranial capacity has more than doubled in less than 2 million years, while certain "living fossils" such as the famous Coelacanth seem to have maintained exactly the same shape for 200 million years or the lingula for 500 million years.

The "objects" whose evolution we are trying to understand—molecular structures, on the one hand and shapes, on the other—are therefore very different; but more different still are the ways of posing the problem. It is only by playing on the ambiguity of words that a false impression of parallel approaches is created: in the case of molecular structures, we are studying a trait so closely linked to the genes that, in general, we have a very poor idea of its influence on the individual in whom it is expressed; in the case of phenotypic traits, we can measure the influence of each modality on the performance of the individual; on the other hand, its link with the genetic heritage, and therefore the mechanism governing its transmission, is, in general, very poorly understood. It is not surprising that the explanatory parameters are, in the first case, linked to drift and to mutations and, in the second, to natural selection.

THE STANDARD NEO-DARWINIAN MODEL

In a first phase, neo-Darwinism developed a model linking the modalities of a phenotypic trait to genes occupying a single locus, that is, in each individual, to a single pair of genes. It is difficult to extend this model to a group of several loci unless one makes the most unrealistic hypothesis that the effects of the various genes on this trait are additive.

Using this model, it is possible to define, from the selective values associated with the various modalities of the trait, the selective values of the various alleles present in these loci and to predict the evolution of their frequencies. This evolution ceases when the alleles have selective values that are all equal; equilibrium is therefore usually achieved by the disappearance from each locus of all the alleles except one, the "best," that with the highest selective value. The effect of selective pressure is therefore to homogenize the population. In certain cases, however, this effect may be to maintain a lasting polymorphism, as when the selective value of heterozygotes is higher than that of homozygotes, or when these selective values are decreasing functions of the frequencies (in this case, the rarer a trait is the more advantaged it is, which is implied in particular by the various models of selection by competition).

Naturally, this theory also takes equal account of the effect, along with natural selection, of the mutations and random fluctuations in frequencies due to the limitation of population sizes. The integration of all these factors can be achieved in a particularly satisfactory way by using Kolmogorov's famous "backward and forward diffusion equations" (which physicists use under the name Fokker-Planck equation). One of the important results obtained (by M. Kimura in 1962) through the use of this formalism is the calculation of the probability of fixation of a new allele produced by mutation in a population, this fixation being the very process by which a species evolves. One finds (see box 5) that this probability depends both on the size of the population that is regulating the extent of the role played by "chance" and on the selective effect linked to this gene. This probability remains low, even for favorable genes; most new genes disappear; nature rarely achieves the "transformation" that it attempts.

The description of the evolution of genic frequencies supplied by this theory seemed satisfactory for a long time, all the more so since it confirmed Darwin's global vision. The most Darwinian result of neo-Darwinism is no doubt Fisher's 1930 "fundamental theorem of natural selection" which we discussed in chapter 2. It states that the average selective value of a population increases at a rate proportional to the variance of individual selective values; global improvement is obtained by the diversity of fates endured by individuals.

The chief difficulties encountered by neo-Darwinism came from the necessity for accommodating the obvious polymorphism of natural populations. In essence, Darwinian selection is "purifying"—the bad are ousted by the good; in the long run, only favorable traits survive. However, nature shows us that homogeneity is the exception, diversity the rule. To be sure, as we have already seen, various modes of

Box 5

Probability of Fixation
of a New Gene Produced by Mutation

Suppose there is a population with, at each generation, N individuals. A new gene a appears by mutation, its initial frequency is therefore $f_0 = 1/(2N)$.

The fate of this new gene depends on its effect on the survival and procreation capacity of the individuals who are endowed with it, that is, on their "selective value"; let s be the difference in selective value linked to a (s is positive in the case of selective advantage, negative in the case of disadvantage).

Usually this fate will be elimination, but sometimes, by chance or thanks to its proper effect, the frequency of a increases; gradually it spreads throughout the population and eliminates all the others; when its frequency reaches 1, it alone remains and becomes "fixed."

It can be shown that the probability F of this fixation is given by

$$F = \frac{1 - e^{-2s}}{1 - e^{-4Ns}}.$$

Hence, when s is very close to zero, $F \cong 1/(2N)$.

When N is very large

$$F \cong 2s \text{ if } s \text{ is positive,}$$
$$F \cong 0 \text{ if } s \text{ is negative or zero.}$$

In other words, only those genes that confer some advantage have a chance of becoming fixed.

However, when N is small, fixation can occur even for unfavorable genes; in the admittedly rather extreme case where $N = 20$, one obtains

$$F = 3.6\% \text{ for } s = +1\%,$$
$$F = 1.6\% \text{ for } s = -1\%.$$

Remember that even favorable genes have only a very slight chance of being permanently retained by the population's genetic heritage.

selection preserve polymorphism, but they imply an unbearable *genetic burden* for a population: they assume in effect that most individuals enjoy a level of viability and fertility that is very much lower than that corresponding to the ideal genotype.

To solve this paradox, the neutralist theory contents itself with not introducing the concept of selection, but it was able to develop only by taking molecular evolution as its subject; it has scarcely any explanatory power when the problem is the evolution of phenotypes. Another path can be explored: making the standard neo-Darwinian model more complex.

COMPLEMENTS TO THE NEO-DARWINIAN MODEL

That on which selective operates is neither a gene nor a pair of genes, but an individual, the result of the interaction between, on the one hand, a collection of some tens of thousands of genes, and, on the other, an environment. In order to be somewhat less unrealistic than the initial models, which linked selective value to a single locus, it is necessary to take account of these interactions by considering, in a first phase, a group of two loci; totally new phenomena appear.

The evolution of the genic structure of a locus therefore depends not only on the proper effects of the genes occupying it, but on the structure of the other locus that is being considered simultaneously. Even extremely simple models, introducing the smallest possible number of parameters, bring to light some unexpected evolutions: a particular gene that is eliminated by natural selection in one population may very well become widespread, because of this same selection, in a neighboring population living in an identical environment.

The mathematical treatment of such processes quickly becomes inextricable; use of computer simulations becomes obligatory. These bring to light particularly the influence of a parameter which has nothing to do with the proper role of the genes: the rate of recombinations between the loci being considered. Once this rate is sufficiently low, linkage disequilibria (see box 6) are maintained, which make all the arguments developed with regard to a single locus irrelevant.

One of the important consequences is that the "fundamental theorem," which seemed to have tied Mendelism and Darwinism into a secure package, is shown to be quite simply false: the effect of natural selection may very well be the reduction of the average selective value of the population.

The study of sets of greater numbers of loci also brings some unexpected phenomena to light: Lewontin has thus demonstrated that, in a series of loci situated side by side on a segment of chromosome and subject to the same selective pressures, the evolution of loci situated at the center of the segment is different from that of loci near the ends; this "embedding effect" may be complicated by the presence of loci subject to strong selective pressures, which draw neighboring loci into their evolution. Certain American authors use the colorful expression *hitchhiking* to denote this drawing effect. It is as though loci occupied by neutral alleles, or almost unaffected by selective pressures, left the business of guiding their evolution to loci that are subject to more rigorous determinisms.

Numerous attempts have been made to specify the evolution of the frequencies in various particular cases. The global result that seems most important to me is that the evolution of a locus, and therefore that of the trait or traits to which it is linked, is a function, in part, of factors that are totally independent of it; the result of the interwining of the mechanisms involved cannot be predicted without complete knowledge of the initial conditions; however, this complete knowledge is obviously impossible. In practice, it is therefore as though "necessity" had left the way open for "chance," provided this word is used as Augustin Cournot defined it: the meeting of independent series of causes.

Box 6
Linkage Disequilibrium

The genes governing the various elementary traits occupy specific positions on the chromosomes, the loci. When two loci are close to one another on the same chromosome, the transmission of the genes at these two loci from parents to offspring is not independent. Imagine an individual with an *A* gene at the first locus and a *B* gene at the second, both from his father, and an *a* and *b* gene from his mother. To produce a gamete (spermatozoon or egg), he makes a copy of the chromosomes he has received and thus transmits either *A* and *B* or *a* and *b*.

However, in the course of the copying process, an event called "crossing-over" (where a break occurs between the two loci being considered) can recombine the genes at the two loci; the gametes then receive either *A* and *b*, or *a* and *B*. This unpredictable event can be characterized by its probability *r*. The closer the two loci, the smaller *r* is.

Within a population, the effect of these recombinations is, in the long run, to make the structures of these two loci independent, to achieve a balance such that the frequency f_{AB} of the occurrence of the pair *AB* in a gamete is equal to the product of the frequencies of these two alleles. When this equality is not achieved, we say that there is "linkage disequilibrium"; its measurement is, be definition

$$D_{AB} = f_{AB} - P_A P_B.$$

In the absence of selection, this disequilibrium tends toward zero all the faster as the rate of recombination *r* is higher. But the effect of selection may be to maintain it permanently.

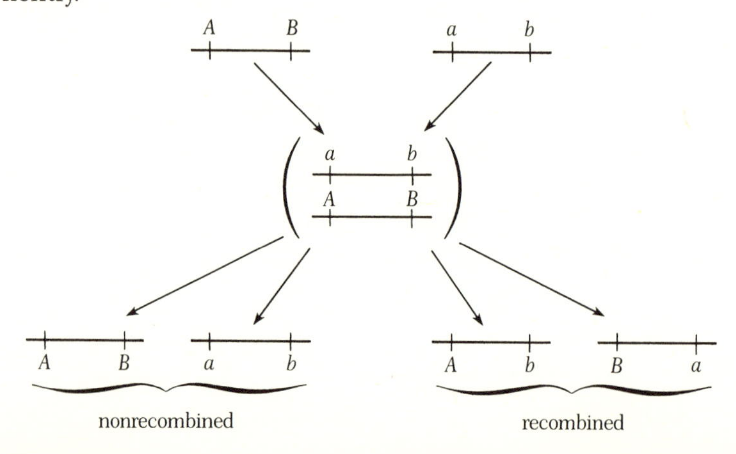

AND HOWEVER . . .

The current summing up may seem very disappointing:

—For certain traits, it is possible to precisely describe the transmission process between generations and to imagine models taking account of their evolution, but these are traits outside of the direct reach of our senses, for which we have no paleontological data and which are, very often, insignificant.

—For others, there is clear proof of their link with the capacity for adaptation to the environment and for success in the struggle for survival, but it is impossible to specify their mode of transmission and therefore to propose a model taking account of their evolution.

This split becomes obvious when current literature in this field is examined:

—Geneticists are developing an increasingly subtle mathematical arsenal, but they do not know how to confront the parameters they have introduced with data from the real world.

—Zoologists and paleontologists minutely describe the transformation of sizes, forms, and colors, but they do not know how to explain the transmission of these traits. In essence, their arguments are based on the gradual adoption, by each species, of traits allowing it to adapt in the best way possible to its environment. Relying on this explanation is playing with words; we discussed how traits, whether innate or acquired, are not transmitted during the process of procreation; only the genes governing these traits are transmittable. Beyond the words being used, it is the very object of a theory of evolution that is ambiguous: for some it is the explanation of a change in the genic structures, for others the explanation of the change in forms and performances.

In this current state of disarray, the clearest need is for new unifying concepts. Lewontin remarks how envious biologists can be of physicists who, by means of three concepts (volume, pressure, temperature), are capable of describing complex phenomena by a formula as simple as $PV/T = cte$. But these concepts, no matter how obvious they may seem to us, were the result of a long process of maturation: for example, two centuries were required to define the content of the word "energy." Population genetics is very young; it is not abnormal that it has not yet fully developed its arsenal of concepts.

New perspectives may undoubtedly be provided by the analysis of the mechanisms of ontogenesis: every individual is constructed based on the input of information, of matter, and of energy from his genetic heritage and from his relatives—his mother, in particular, during gestation. The geneticist focuses his attention on the genetic heritage and tries to explain the apparent changes during the evolution of a line by the modifications to this heritage. The zoologist focuses on the relationship between the trait expressed and the environment; he observes an adaptation of the former to the latter, but does not know along what lines the genetic heritage could have been modified in order to achieve this adaptation.

Recent discoveries on the noncolinearity of proteins and DNA only make this problem more difficult: it was still believed up until a few years ago that the sequence of amino acids constituting a polypeptide chain corresponded exactly to the sequence of DNA bases; it is now known that this process includes a splicing phase that puts noncontiguous regions end to end. Thus, the hemoglobin B chain is synthesized not from a single segment of DNA, but from three segments, each providing the information necessary for synthesizing a part of the chain. The production mechanism is therefore much more complex than had been imagined; it involves phases of splicing and rejoining of segments with extreme precision. This mechanism is obviously itself dependent on the genetic information. The structure of this hemoglobin chain thus depends not on one gene, but on several.

Now the theory of evolution assumes that the final target of natural selection is, through the individual, the gene responsible for a particular structure. How is this gene to be reached if the final product results from the combined interaction of multiple causes?

Here no doubt lies the core of the difficulties: Mendelian genetics was built entirely on the equivalence *1 locus* ⟷ *1 trait* (for instance, the wrinkled surface of Mendel's peas): in this model, a new gene appearing at a locus can modify the trait; natural selection then affects the carrier and, through him, the gene in his descendants. An arrow describing the retroaction of the environment on the genetic heritage can therefore be drawn: this arrow is the very basis for the neo-Darwinian theory.

We now know that this simplicity is rarely encountered: a locus may influence several traits; any trait, even one as elementary as the structure of a hemoglobin chain, depends on several loci. An entirely different equivalence has to be taken into account: *several loci* ⟷ *several traits*. And we no longer know how to draw the arrow describing the effect of the environment on the genetic heritage.

To extricate ourselves from these difficulties, we resort to describing statistical liaisons and to using the probabilistic argument. Is this a temporary stopgap or a permanent solution because it is in harmony with the very nature of the problem? The reply depends on one's personal sensibilities.

Note in conclusion that this reply has far-reaching implications for our way of looking at the living world: depending on whether this world is the result of a strict determinism that led it in the only direction possible, shaped it in a way that it could not fail to be, or on the contrary, the end product of a partly erratic process, which, under the same conditions, could have forked in entirely different directions, our feeling about the place that we humans occupy in this living world and the role that we can play is profoundly modified.

It is not immaterial to know that, in its current state, science does not answer this question.

NEW DEPARTURES

Every scientific conference evokes the seriousness and the imminence of the dangers resulting from humanity's current dynamic. It seems odd to hear so many rigorous minds declare that disaster is certain to happen any day, and then continue with the day's agenda. This ship that is rushing toward the rocks, would it not be useful to change its course, rather than continuing with our laughable distractions? According to legend, Byzantine intellectuals continued their exciting debates about the sex of angels, while the Turkish armies were, before their very eyes, preparing the final attack. Are we going to continue to increase the efficacity of the structures and machines that bring the end ever closer, and leave lucidity to a few powerless Cassandras whose predictions produce scarcely any effect other than a delightful tremor of anguish that makes the comfort of the present moment even more precious?

Slow readjustments to the current human dynamic are likely to be useless; the effects would be felt only in the long term—it will be too late. Drastic changes, whether planned or endured, will necessarily occur.

Extrapolation from current trends leads only to "scenarios of the impossible." The physical limits of our planet do not allow us to envisage, for the second half of the next century (so close! our grandchildren will be scarcely any older than we are now), a world population of 12 billion people, enjoying the same nutritional and energy possibilities as the average French person of today. Social limits will make themselves felt even sooner: the foreseeable rates of development will lead, in the year 2000—that is, tomorrow—to a world distribution where two thirds of all riches produced will be attributed to the 1.2 billion inhabitants of industrialized countries (Europe, the Soviet Union, Japan, North America) and a quarter to the 1.8 billion inhabitants of countries that have just begun industrialization (for example, Mexico, Brazil, China); as for the remaining 3 billion people, the "underdeveloped," they will receive a twelfth of the total, while they will represent half the world's population (estimates by A. Danzin, quoted by J.-M. Legay).[1] For how long will a difference in standard of living of between 1 and 30 or 40 (not to even mention the infinitely greater individual differences) be tolerated by those who have massive numerical superiority over the wealthy and powerful?

The actual scenario will inevitably include "catastrophies" as defined by mathematicians, that is, sudden changes in certain characteristics of our societies. Instead of awaiting them passively, is it not possible to anticipate them and prepare some new departures that will forestall the ukases imposed by necessity and that will allow us to maintain a certain mastery over our destiny?

A few months before May 1968, Robert Mallet recalled Rivarol's advice to the Prince: "If you wish to avoid revolution, make one!" Thanks to our wonderful intellectual apparatus, we are capable of anticipating events, of describing the coming catastrophe before it happens and therefore of preventing it. It is now urgent to use this capacity to imagine the obstacles ahead of us and, if possible, to avoid them.

In this research, we must be careful not to give preference to our own point of view as a French, European, white or rich person; we must, in so far as possible, try to think as "Eartheans." Humanity is actually undergoing before our eyes a mutation similar to that of a solution that suddenly crystallizes, or that of a liquid which, because of the effects of a decrease in temperature, changes to a solid. For a long period of time, the various human groups scattered around the planet had only rather loose links with each other; they were still, a little while ago, almost independent of each other. The development of communications and the more far-reaching impact of our actions have created an interconnectedness which makes each individual dependent on all the others: the desert nomad is informed by his transistor of the latest quarrel between oil producers and purchasers; the trapper isolated in the enormous frozen wilderness of Labrador receives rain made acidic by smoke from industries located thousands of kilometers away. There are no longer any inaccessible retreats where we could find shelter from the actions of others. A global rigidity of the whole of our planet has appeared. Whether we like it or not, we are interdependent; the fate of all humanity is at stake, and one person's choices are liable to affect the future of all. Trying to extricate oneself or one's group as best one can at the expense of others would be but a very poor calculation.

In less than a century, over 12 billion people, but, right now, several thousand atomic bombs stockpiled by powers all wishing ardently, sincerely, for peace, but ready to "press the button" to dissuade the other side from doing so. We must both dread being too numerous on earth and fear ceasing entirely to be.

Our efficacity is almost unlimited when it comes to destruction. If a selective catastrophe were to wipe out all humans, and them only, the other species could breathe a great sigh of relief. For our presence is a threat to all living things, and because of us life may be extinguished.

Bemoaning and decrying the attitudes that have led to this situation are equally useless. An antiscientific attitude, condemning science in the name of the risks it has entailed, is like the ostrich burying its head in the sand so as not to see the danger. Let us repeat once again, science is not one product among others of the human species; it corresponds to a uniquely human attitude to the universe. Renouncing the lucidity, and secondarily the efficacity, that science seeks would be to renounce our human specificity.

Is it possible to imagine a realistic path that would enable us to avoid the dangers without losing anything essential?

Confronted with a question of this kind, the best we can do is to propose some threads of thought. The preceding chapters have attempted to describe, in some sectors of biology, scientific activity as it is experienced by those who are engaged in it: a trial and error process, involving the reexamination of existing concepts and the laborious elaboration of new ones better adapted to reality.

The dangers facing humanity have reached their current gravity only because of scientific activity; the study of the status of this activity in our society is therefore urgent. In the pages to follow, we will outline, without logical or hierarchical order (this could only be a delusion) some tentative approaches, some basic reexaminations, some lines of exploratory research, which could contribute to putting scientific activity on the right track.

1. Human Rights

Human beings, all of whom are different because of the mechanism of sexual reproduction which mixes parental genes to create endless new combinations, are gradually shaped by personal experiences, which accentuate these differences even more. Putting an "equal" symbol between two people is, quite obviously, impossible. But—we stressed this when dealing with the "numbers pitfall"—neither can they be classified according to a scale of values, from less good to better.

The equality that features on the pediments of our monuments is not equality of individuals, which would be meaningless, but equality of the rights which society grants them. This is a neat formula, it is quickly said, but what exactly does it imply?

It is nearly obvious that the most immediately efficient social organization is of a kind that we call "elitist": 1 or 2 percent of the population are declared worthy of governing, 4 or 5 percent are responsible for the maintenance of order, and the remainder carry out the various tasks necessary to the survival and well-being of the group, under orders from the first two. "Citizens—soldiers—slaves," "*alpha—gamma—epsilon*" terminologies vary according to culture or author, but the structure is the same: a small number decide on the objectives and issue orders, a slightly larger number transmit the orders and supervise, the immense majority obey and produce.

The remarkable efficiency of this structure is proved, for example, by the remains of the Greek cities. How inspiring it is to visit the ruins of Syracuse, of Agrigente, or of Selinonte, to imagine the splendor of these cities, protected by enormous ramparts, adorned by temples of unsurpassable beauty, where it must have been so pleasant to converse with brilliant mathematicians or profound philosophers, while contemplating the sunset on the sea!

But each of these massive stones, now crumbled, must have been extracted from a quarry that was sometimes quite a distance away, cut, carried, put in place; the lives of thousands of slaves were devoted for centuries to the building of these perfect cities. Which memory should be associated with these ruins—the harmony of the citizens' lives or the sweat, blood, and death of the slaves?

Restricting all rights, of decision making and of enjoyment, to a limited "elite" is usually a guarantee of efficiency. This efficiency is due to the very structure of the society and not to the personal qualities of the members of this elite. Whether these be chosen because of their ancestors (hereditary nobility), of their height (as, for instance, among the Tutsi in Ruanda), of their skin color (as in the American slave states in the nineteenth century or in South Africa today), or of pure chance (as in the case of the Dalai Lama in Tibet) is of no importance; what matters is that they are few in number, convinced of their "natural" vocation to lead, and capable of convincing others of this.

Unfortunately, a day comes when this ability to convince others declines. A subtle poison contaminates the elite's belief in the legitimacy of its power; subversive ideas, disseminated by the disciples of a Jesus or by the readers of a Rousseau, spread among the *epsilon*. They are no longer content with the lot, however comfortable, that is assigned to them; they demand "equality" of status with those in power.

Questions about whether their demand will have favorable or unfavorable consequences for the development of society, or for their own material well-being, are irrelevant. The efficiency of an elitist structure is perhaps so great that, even when frustrated, the *epsilon* may attain greater wealth than within a structure recognizing the equality of their rights. But the question is not put in these terms.

What is involved is the definition of our concept of "otherness" and, at the same time, of our concept of self. Are we just cogs in an enormous machine, society, whose smooth operation is the ultimate goal of the group? Do we, on the contrary, represent the supreme value (as a "child of God" or as "an individual equal in rights with all other individuals"), whose fulfillment it is society's role to promote? Our reply to these questions implies one of the following: a society at the service of individuals or individuals at the service of society?

This reply is not provided by a scientific argument: it can only be asserted. Even then, it must be clearly posited and the consequences drawn without cheating.

Let us listen to all the speeches. At the present time, there is perfect unanimity: man is the supreme value, and society exists only to allow him to develop, while respecting the rights of others; each person's rights are imprescriptible; there is a plethora of slogans and of speeches at every meeting, and declarations are made constantly about "human rights"; the UN adapts these declarations, and all states solemnly subscribe to them.

But look at the reality: what is the weight of the billions of people who laboriously eke out a scanty livelihood compared to the few peoples who have appropriated the wealth of the planet? Within these very peoples, what is the weight of the millions stuck in the insipid daily round of imposed work, of information and entertainment passively absorbed, compared to the few "princes" who govern them?

The "rights" so generously ascribed by the United Nations charter or by state constitutions are, in practice, scarcely ever granted. Perhaps this is necessary for the sake of the economy, or of security, but then why prolong the hypocrisy?

Here words play their most pernicious role: they contribute to justifying and perpetuating a reality that is the opposite to the reality that they express.

If ducks could understand our language, we would succeed not only in persuading them that a stuffed duck is happier than a wild one, obliged to search for its food, but also in making them proud of the monstrous livers that our gluttony obliges them to endure. The endless statements about human rights have succeeded in persuading people to relinquish their rights in order to defend them better, in making them proud of the wounds they have suffered while contributing to the oppression of their fellow humans.

In our own so-called developed societies, the mockery is simply too much to bear: technological progress has been such that our capacity for producing riches is superabundant; the efficiency of an elitist structure, which could, in former times, be evoked as justification for inequality of rights, has lost its relevance—a machine can easily replace a thousand slaves. Something that was once a luxury which society could not afford is now within arm's reach, but, out of fear of having to undertake too fundamental a reassessment, we leave the old order in place, when its justification has disappeared.

In just a few hours, each of us creates more riches than our ancestors working to the limit of their endurance. The spread of computer science will lead to a new leap forward, but instead of being delighted, we are worried, because we see it as another source of unemployment. Is it not time to at least draw the consequences of this progress by organizing a society where everyone's rights would be equalized?

We see, quite to the contrary, an exacerbation of competition, of selection, and of the elimination of the greater number to the benefit of the few, while declarations about human rights are more numerous than ever.

This equalization of rights has nothing to do with a general leveling, a mass production of individuals all in conformity with an imposed model. On the contrary, it involves fostering the development of individuals in all their diversity, respecting differences, that is, nonequalities, without attaching a value judgment to the latter. It is easy to label those who strive for equal rights "equalitarian fanatics" leading to a destructive leveling; this is a misappropriation of meaning as despicable as a misappropriation of funds. These abuses of language, which confuse so many readers, indicate that the struggle for "human rights" begins with the rigorous definition of words, because words are weapons.

If we are really committed to human rights, they can become as valid a subject of scientific research as quarks or quasars.[2] Let us try to imagine that, throughout the planet, the oft-proclaimed equality of rights is actually implemented; all the knowledge of economists, sociologists, jurists, and many others will not be too much to try to predict the consequences of this change.

2. *Professionalism and Deprofessionalization*

In the course of their development, our western societies have attained an increasingly high level of professionalization and specialization. All areas of human activity have been marked by this trend; at a very early stage of a child's development, there is a concerted effort to direct him toward a well-defined intellectual or manual activity. In order to fit into society, to play a role within it, to obtain from it what we need for our survival and development, we have to acquire competence in a specific field of activity—in other words, to become a professional.

In chapter 1, we emphasized the dangers of isolation, of frustration, and even of mutilation of individuals, of the compartmentalization and sclerosis of society engendered by this process. The advantages are obvious; it is the distribution of tasks and the specialization thus made possible that have brought about increased efficiency and driven the age-old anguish linked to hunger, cold, and disease far from us.

This specialization seems especially necessary in the scientific domain, so rapidly is knowledge increasing, but it is also especially dangerous because of the gap it creates between disciplines, and especially between the scientific community as a whole and public opinion. How can we try to eliminate these harmful consequences without renouncing the advantages of professionalization?

AN EXAMPLE FROM ARCHITECTURE

Before examining what could be attempted in order to deprofessionalize scientists, let us consider an example where the simultaneous struggle against the impoverishment due to deprofessionalization and the decline due to professionalization achieved a degree of success. This happened in architecture.

The construction of a building as complex as a high school for 1,500 to 2,000 students obviously requires input from a "professional" architect whose competence is absolutely necessary if one wishes to respect safety regulations, to follow the guidelines laid down by the Ministry of Education, to ensure traffic flow, and even to end up with a building that is aesthetically pleasing (which rarely happens).

For the first time, at least in France, the architect responsible for drawing up the plans for such a high school in a large city in the west of the country refused to submit the project to which his reflections, his analysis of the program, and his critical examination of previous projects would have led him. Yona Friedman[3] felt that his role was not to apply his own knowledge, but to share it with the future users of the high school, who thus became the real architects. He wrote a "manual," a kind of comic strip that was a rigorous summing up of his courses (at MIT); this "manual" explained the most advanced methods in the field of architectural design to non-experts. The future "autoplanners" were thus better informed than many professionals, thanks to the information thus made available to them, that is, formulated in their own language. Then he called a meeting of all those administrators, teachers, students, and parents who were interested in the project and asked them to draw the plans based on the imposed program; his role was not to criticize the options chosen, or to sway the decisions that gradually emerged, but simply to check their coherence and conformity to administrative, financial, and technical imperatives. After an inevitable period of flux, a group dynamic gradually developed, and agreement was reached on the principal choices and on the articulation of the various elements of the future structure; a plan was adopted; construction is in progress.

One incident proved the effectiveness of the process thus established. The future high school bears little resemblance to the many all too similar ones that have been built in France over the past few years. Its unusual design disconcerted some members of the administration, who attempted to have the entire project reexamined; the reaction of the architect-users was human—the common front (which included all the autoplanners, whether on the "right" or on the "left," a unity of action that has scarcely any precedent) made it possible to overturn this opposition with a ministerial decision. The new high school will indeed be "their" school.

This example seems to me to be indicative of an attitude that puts the problem of deprofessionalization in a new perspective.

In this case, the professional architect did not in the least abdicate his role; he did not hide his competence, his knowledge, his experience, and his know-how under a bushel. He knew on the contrary how to shed light on them so that they could be understood and shared. Above all, he assumed that his role was not that of a guide, but that of a stimulator to self-guidance. To use a biological image, he was neither the brain that reflects and decides, nor the eye that sees, nor the muscle that acts, but the endocrine gland that spreads a certain "way of being" throughout the organism. He did not "deprofessionalize" himself—he kept all the richness of his "professionalism," but he knew how to escape from the prison of unshared competence; he lived a collective adventure where this richness could act in synergy with the riches provided by others.

THE PROFESSION OF SCIENTIST

Can this one isolated experience be transposed to other domains, and if so under what conditions? It would be an illusion to try to find a rigorous reply to this question; let us try to specify it in the case of scientific activity.

The subgroup of our societies which the scientific community represents is only one special case, but a very interesting case for our subject, because the process of enrichment-decline due to professionalization is easy to characterize there. The questions that can be asked about this subject have a clear meaning when the specimen singled out for observation is a "scholar"; what has his specialization, which is necessary for him to truly play his role, done to him? What use can he make of the power which his halo of lay saint confers on him? How is he manipulated by others? To what extent is he the product (the subproduct) of a society that secretes knowledge just as it secretes gadgets? To what extent is he, on the contrary, the product of a personal "vocation"? Any or all of these questions could lead to significant explorations.

Two or three centuries ago it was possible to become a universal scholar, up to date with knowledge in all domains, whether astronomy or chemistry, botany or medicine. The scientific explosion, due above all to the development of increasingly accurate tools of measurement, brought about a compartmentalization of knowledge. Even within a discipline with relatively defined limits, say, biology, no one can really keep up with the latest hypotheses or the new findings whose results are published in countless specialized journals. The careful reading of just those few journals that reach me regularly would require much more than my entire work time. Gradually, out of necessity, everyone takes refuge in an ever narrower branch of the tree of knowledge. This is necessary not only for one's personal intellectual comfort, but also for efficiency.

Genetics is but a branch of biology, but it is itself already too ramified for it to be possible for one person to master all its aspects; keeping up, for example, with the development of "population genetics" alone is the most one can reasonably hope for. A genetics congress has become a Tower of Babel where the various sections are united by scarcely anything other than the word "gene," which moreover does not have the same meaning for those involved in molecular genetics and those in quantitative genetics. Unity of interest and of language is to be found only in conferences or seminars with a limited objective where a few specialists co-opt each other and meet to pool their insights.

When I look back on meetings that were of great benefit to me and that had a lasting effect on my activities, I do not recall a single major congress, but, instead, a meeting in Israel of about fifty population geneticists studying for two weeks the problems posed by the elaboration of mathematical models, or, in Iowa, of most of the authors of articles and books on "heritability," in order to reflect on the difficulties associated with this concept. While large congresses are now little more than futile society gatherings, symposia with limited objectives and numbers are fruitful for the development of one's thinking. Thus, little by little, serious scientific activities are becoming confined to highly specialized groups of "professionals."

This judgment is, I imagine, valid for most disciplines. A scientist is becoming similar to those radar beams which reach all the farther the narrower they are. He is typically a "specialist." But the obvious efficacity of science as a whole gives the researcher an authority to which the public responds spontaneously. He is merely radar with a narrow beam, but he is expected to be a lighthouse sweeping the horizon.

Solicited to give his opinion and to intervene in various domains, he does so willingly even when the domain in question is scarcely touched on by his specialty. This can be seen as a veritable abuse of trust or a usurpation of authority. But can the scientist refuse in such cases, and would he be right in so doing?

THE SCIENTIST "DEPROFESSIONALIZED"

Unless he is completely "polarized," blinded by his subject, the specialist is led to draw certain consequences from the lessons provided him by his research. In the case of population genetics, this fact is especially obvious. Thinking about the process of renewal of living beings, about the mechanisms involved, necessarily modifies the way we look at others, at society, at ourselves. Most of the notions implicitly accepted by our societies—determinism of traits, classification of races, hierarchy among individuals—are brought into question. The obvious misinterpretations flawing most arguments developed on this subject and the errors in behavior which they justify lead the scientist to intervene, to get involved in polemics, then, gradually, in political issues. In so doing, he does of course become deprofessionalized, but it seems to me that he can do this without damaging his specialization, and that he can even enrich it, by validating it.

For what motivates his involvement is not the store of facts that he has accumulated (through his own observations or through reading), it is not the detail of mathematical developments that allowed his thinking to progress; rather it is the personality structure that he gradually acquired from these data and this thinking. However, this structure can be achieved in a much simpler and more easily transmissible way than by following the necessarily tortuous, uncertain, and aleatory personal development of which it is the end result.

Édouard Herriot's famous quip about culture is worth quoting here: "what is left, when everything else has been forgotten." It is not really a matter of forgetting everything, but of reconstructing and rearranging the set of insights which we have developed. Once this work is done, the transmission can be done without professional jargon; the effort required to understand an idea is minimal compared to the effort that was required to hammer it out. The professional is therefore like the first person on the rope, who, after endless trouble and dangers, reaches the summit of the peak and throws the rope to his comrades who can then hoist themselves up without much effort.

In the scientific domain, concepts function as a rope. Every discipline advances laboriously by trying to take account at each moment of all the available facts; according as these accumulate, the theories and explanatory models become increasingly complex, within reach only of specialists and professionals, until a unifying or simplifying concept is proposed, which permits an easily transmissible global understanding. Let us evoke three examples: the clearest is no doubt that of *universal gravity,* a concept that, in spite of its abstraction and blatant unreality, makes it possible to easily explain the falling of bodies and the movement of the planets. More important is the concept of *chance,* as it is defined in the probabilistic argument; it alone makes it possible to develop a realistic attitude in the face of a necessarily uncertain and possibly undecided universe. Less widely used at present, but likely to become increasingly popular since it enables us to ask better questions, is the concept of *undecidability,* which makes us conscious of the limits of rigor.

It is only after lengthy explorations that these notions could be developed and gradually clarified and that their relevance to scientific practice (whose ultimate objective, remember, is not effective manipulation of the universe that surrounds us, but coherence in our representation of this universe) could be established. But once this time-consuming work has been accomplished, its result can be comprehended with minimal effort. The scientist must, in this phase, without losing anything of his specialization, put the conceptual tool that he has forged at the disposition of others; this tool can then be used even by those who do not have all the knowledge that had to be accumulated to forge it. (Even then, they should not relax their vigilance as to the pertinence of the concepts and terms thus used. Let us repeat, in our exploration of the universe, the essential tool is the word; a poorly defined or understood word is more dangerous than a jagged scalpel.)

The scientist is here in the same situation as the architect mentioned earlier who did not create for others but created circumstances in which others could create (understanding is as important for each of us as loving; it is an activity that cannot be delegated; we do not put Cassanova in charge of our love life—let us not put scientists in charge of understanding for us).

In applying himself to this diffusion of explanatory models rather than of facts, the scientist will certainly lose his status as mysterious and all-powerful "man of learning," but he will really do his job. He will play his role by putting society as a whole in a position to control him.

In order to achieve this necessary change, it is the entire educational system (and no doubt also the status of researchers) that must be reconsidered.

3. *The Educational System*

Far better than abstract considerations, the recounting of a few personal memories will, it seems to me, contribute to clarifying the problem posed by the insertion of an educational system in our society.

UNUSUAL LEARNING EXPERIENCES

Dr. F., coordinator of the "People's University" at Moyeuvre-Grande, asked me to moderate an evening devoted to genetics. On picking me up at Metz, he is worried; in this small city in industrial Lorraine, how many people will be brave enough this evening to listen to a presentation on such an arid subject? Thirty or forty, he hopes.

When we reach the town hall square, a crowd carrying chairs is on its way toward the banqueting hall, which had to be opened and hastily fitted with a public address system; more than 500 people, workers, teachers, and pupils have come. For over three hours, after a quick presentation, question follows upon question, all of them pertinent: we discuss serious matters, races, defects, and biological fatality. Does scientific fact justify—as a strong current of opinion, so well underscored by the mass media, attempts to make people believe—brutal selection, oppression, contempt, hierarchy among people? In spite of some necessarily rather technical explanations, the audience's interest never flags; it is very late when we separate.

On the invitation of science, mathematics, and philosophy teachers, I spent an afternoon with a group of Grade 13 ("terminales") students in a high school in an eastern suburb of Paris; the audience is larger than expected, because the Grade 12s ("premières") skipped their classes to take part in the debate. For want of a better place, we pile into the refectory; the acoustics are deplorable, but what does it matter? There is no letup in attention all afternoon: we discuss the "making" of a child, not, to be sure, the necessary physical acts, which they already know, but what really happens—what is transmitted? How does one become what one is? What is innate, acquired? Once again, questions that are vital to all of us—determinism, fatality, freedom—are seriously evoked. I have not forgotten the remark made to me by one of the teachers after the meeting: "These kinds of meetings should take place more often. So many of our students, when confronted with failure, give up the struggle completely, accept their predestined lot as unskilled workers and say to us: 'I'm worth nothing'"

It is terrible when education, which should help people to grow to their full potential, actually leads them to despair and to feel totally worthless.

The pirates at "Radio-Lorraine-Coeur-d'Acier" decided to devote most of the evening program to genetics one Thursday in June 1979. They asked me to participate. At about eight o'clock, the program begins. In more than half the households, for fifty to a hundred kilometers round about, depending on the whims of wave propagation, the television set is turned off and the "local radio" is turned on. A jamming transmitter, installed at the television relay that dominates Longwy, tries to make LCA inaudible; however, makeshift antennae have been set up everywhere; the local technicians have been initiated to tuning "in counter phase," and finally LCA can be heard loud and clear, just like the BBC of long ago.

A plank hut in the foyer of the town hall is used as a broadcasting studio; the overcrowding, the disorder, the constant comings and goings come as a surprise to someone accustomed to "official" radio stations. Here, the door is constantly open; anyone who wishes can come in, listen, express himself. Crowded around a table covered with microphones are two journalists, Marcel and Jacques, the only "professionals" involved, two educators, two immigrant workers, a young woman in a wheelchair, a young boy, a few friends who came just because they lived close by, and the guest of the day responsible for answering questions about racism, the heredity of intelligence, and the "highly gifted." Behind the table, the technical console is manned by Leonard: eight hours a day he carries bags of waste in one of the dustiest workshops of the steelworks; after work he acts as radio-technician at LCA, plugging in microphones, regulating volume, and sometimes he also makes announcements on the microphone. His slight stammer does not prevent the message from getting across (on the contrary, "We are the only radio station in the world with an announcer who stammers").

The genetics class begins, quickly interrupted by questions that bring the presentation back to essentials, to matters of everyday living: the contempt to which Arabs and blacks are subjected, the "selection" imposed on children to direct them toward shorter, nonacademic programs, the fatality of an impoverished life; is this fatality biological, "innate," as so many books claim? The debate begins in earnest, between people who know they are involved and are trying to become informed and to understand.

An elderly woman comes in, walks hesitantly around the table a few times, then makes up her mind and unfolds, like a flag, a large embroidered tablecloth. Genetics are forgotten, the embroidery is admired. "I don't know how to do anything on the radio; so, in the course of two months, I embroidered this tablecloth for you. You can auction it off." Then, over the microphone, she explains her embroidery technique.

Another interruption: a teacher, a lover of modern poetry, draws a parallel between what was explained by the geneticist and a recent poem; he comes into the studio and reads a few lines beautifully. Biology, thanks to him, takes on another dimension.

Until 11:30 p.m. genetics, poems, and current events mingle. Then the equipment is tidied up, and the whole team leaves for dinner with friends in their home.

One word characterizes this extraordinary success which goes far beyond the initial objectives: fervor. It is not just a union spreading its slogans, not just a company fighting unemployment; it is an entire population exercising its right and its power to inform and express itself.[4]

On learning about this aspect of my activities, a journalist who does not exactly wish me well introduced me to his readers as someone who "performs in school yards." This was intended as an insult, but is it not on the contrary a compliment?

I have stressed this point repeatedly: at present, communication between scientific researchers and the general public is very poor; this gap could have very serious consequences for our society; to fill it, convergent efforts are required, and first of all those of scientists themselves. In directly addressing those affected by the results of their research and thinking, they are simply doing their job. The obligation to express themselves in plain language, without the help of the scientific jargon that they use with their colleagues, is for them an excellent constraint, an opportunity to better implant their thinking in the real world. The time spent on this "popularization" is not really lost; they are enriched in return by real questions of a kind that most of us encounter in our day-to-day lives, but from which scientists are protected and therefore risk neglecting.

To be sure, it may seem less prestigious to address a group of workers who want to learn about theories of evolution than to teach a course to doctoral students on Kolmogorov's equations, but the long-term impact is certainly greater. Above all, one is responding by so doing to an appetite for knowledge, which seems more obvious to me in certain supposedly "working-class" milieus than in universities.

Of the radical changes that our society needs to make, those involving the educational system are undoubtedly the most urgent.

THE TEACHING MACHINE

This system, by a possibly natural and spontaneous process, has gradually become specialized, "professionalized," and has completely lost sight of its true purpose. Conceived with a view to giving each individual what he needs for his development, it has become a huge machine geared to provide society with individuals ready to carry out the tasks required by its various structures.

It functions like an assembly line as inhuman as that of an automated factory; at each stage, the only objective is to prepare the next stage. The goal is no longer to think, ask questions, and make choices but to absorb what is on the program and to arrive, before the others and armed against them, at the narrow entrance door to the workshop that follows.

The key word has become "selection," that is, the gradual choice of those individuals supposed to constitute the "elite." But for the vast majority this word spells "elimination," that is, rejection. It is as though pupils and students were being injected into an enormous distilling tower that rejects at each of its floors the most volatile products; only those corresponding to a specific qualification are deemed worthy of reaching the next level; the others are directed toward less demanding circuits and to a less promising end.

Administrators, teachers, parents, and students themselves all contribute by their everyday activities to strengthening this mechanism, even while they perceive its absurdity and denounce the detrimental role it plays.

Instead of leading young people to explore ever broader horizons, education actually channels the vast majority of them toward narrow streams and sometimes comfortable dead-end jobs; worst of all, it leads them to accept a blocked destiny. Branded as failures by the mechanism of selection-elimination because of their inadequate "abilities," they accept this rejection and become the willing prey of the herd makers.

Scientists have defended this mechanism and have even found a clever word for it: "meritocracy." To be sure, they argue, the fate of the eliminated is deplorable, "but it cannot be helped; each one is judged according to his abilities and obtains from society opportunities appropriate to these abilities. Nature has given everyone different aptitudes; selection has no function other than to draw the consequences of this fact."

How much truth is there in this argument?

MERITOCRACY: MYTHS AND REALITY

In a book that had a great impact in the United States, the American psychologist Herrnstein states that Western society is tending toward a meritocratic model, where merit is measured by the intelligence quotient.[5] According to him, each person's place in society will be predetermined to an increasing degree by his "innate intelligence," such as it is revealed by IQ. In France, this thesis is restated by the child psychiatrist whom we have quoted several times:

> *Why are there so few workers' children at university? . . . The hereditary character of the intellectual faculties plays its role. It is* natural *that, in a social class where the average IQ is below 100, the number of children fulfilling the requirements for university should be relatively low. This reality must be faced. Deploring it does not change it. And neither does ideologically inflating the effects of the so-called sociocultural handicap.*

I have already discussed how vague the concept of "innate intelligence" or of "intellectual potential" is and have emphasized the difficulty of going from the notion of intelligence to the measure labeled IQ. Without returning to these problems, let us enter into the logic of the IQ meritocracy adepts and try to check its coherence.

This analysis was carried out recently by Michel Schiff, who confronted the meritocratic hypothesis with current reality in France, as represented by the official statistics of the Ministry of Education.[6] These very detailed statistics make it possible to calculate for each "social class," the probability of reaching the various school and university levels; however, even though the data are published, the significant figures that they make it possible to calculate are not.

Thus, in 1978, the number of "children of workers and foremen" who were registered at a university (excluding the institutes of technology and the professional schools) was about 17,000. This number may seem to be very high and to justify statements about advances in the democratization of higher learning. But this social category represents 40 percent of one age category, or about 350,000 young people. The probability of getting accepted to a university can therefore be estimated, for this category, at 17,000/350,000 = 5 percent.

If the same calculation is carried out for the most privileged social category, the children of "senior executives and professional people," this probability is 58 percent, or eleven to twelve times higher. It is obviously these two percentages, 5 percent and 58 percent, that make it possible to argue about the democratization in question, not the absolute numbers.

As a provisional working hypothesis, assume that the findings thus obtained are, among the children of senior executives, the consequence of meritocratic determinism based solely on IQ: those whose IQ is above a certain threshold enter university, those who are classified below it are rejected. It is possible to calculate this threshold thanks to the studies carried out for the past thirty years by the Institut national d'études démographiques on the intellectual level of school-age children. A sample of more than 100,000 schoolchildren ranging in age from six to fourteen was studied. This study made it possible in particular to specify the distribution of children's IQ according to the characteristics of the parents: it was found that, for "executives" as a group, this distribution is in conformity with a "normal law" (which is represented by the classical bell curve), whose average is equal to 111.5, with a standard deviation of 13.6 (which means that two thirds of the scores are situated between 111.5 − 13.6 = 97.9 and 111.5 + 13.6 = 125.1).

In a distribution of this kind, the IQ surpassed by 58 percent of the group (the proportion of executives' children who go to university) is 108.8; according to the meritocratic mechanism, access to universities would therefore be reserved for IQs over 108.8.

However, the same studies show us that children of workers and foremen have a "normal" spread of IQs with 96.7 as average and 13.5 as standard deviation; it is deduced from this that the proportion of them surpassing 108.8 is 19 percent: we saw that in reality the proportion of these children entering universities is only 5 percent.

In other words, the hypothesis than an IQ meritocracy is indeed functioning among the children of senior executives is incompatible with the hypothesis that it is functioning among the children of workers. The number of the latter entering universities should be multiplied by 4 for this type of equality to be achieved.

But the efforts required to achieve the famous "equality of opportunity" that is so often discussed are greater still if one takes account of the effects of improved educational conditions on IQ. In chapter 5 I discussed the results observed on children born to families at the "bottom" of the socioeconomic ladder and reared in families classed at the "top" of this ladder. Even if we assume that the initial handicap shown by these children can only be half compensated for by a better environment (which is a very pessimistic hypothesis, given the findings), the distribution of the IQ of the children of workers and of foremen would be 104.1 on average and the proportion of them surpassing this threshold would be 36 percent.

Let us take a look at the human reality behind these cold figures. Clearly, while at least 36 percent of workers' children could, in an "IQ meritocracy," go on to universities, only 5 percent do so at present, that is, seven times less. In other words, the circumstances under which they develop and under which they are confronted with the various obstacles that punctuate their schooling eliminate, within this social class, 6 children out of 7, without it being possible to invoke any biological fatality whatever. This is not a plea motivated by "egalitarianist" preferences, but the conclusion of rigorous observation. It may surprise, since it goes contrary to so many received ideas that are expressed daily as obvious facts. These prejudices are clearly evident in the conclusions of a former rector, who wrote: "*Even if it is unpleasant to admit it,* the genetic potential *for success is higher, statistically, in the descendents of those individuals who have managed to achieve greater success than others.*"[9] Nobody denies that social success is transmitted from one generation to the next; however, the problem that seems to escape this author is to decide whether the cause of this transmission is "genetic."

These figures give but a very limited idea of the intellectual waste of our society. The scale of IQ scores itself bears the mark of the effects of social discrimination. Relying on statistics from the thirteen OCDE countries, Schiff estimates that from 80 to 90 percent of children whose parents belong to classes on the lower half of the social ladder are excluded from university studies for reasons other than their IQs.

These results show the unimportance of questions about heritability: "Wondering whether genetic differences between social groups explain 0, 50, or 100 percent of IQ differences amounts to wondering whether the waste of intellectual potential is 97, 94 or 'only' 87 percent."[10] They show above all the importance and the urgency of the efforts to be made in light of a true "ecology of the spirit."[11]

"DESCHOOLING SOCIETY"?

Analyzing the detrimental role that the educational system can play, Ivan Illich provocatively titled his book *Deschooling Society.*[12] To be sure, school such as it is structured contributes to inequality of opportunity, but it is only one element in a much larger structure.

What is at stake is each person's right to have access to knowledge. For this right to be respected and implemented to the same extent as a sick person's right to treatment, the entire society has to become involved. Rather than deschooling society, it is a society where everything is school that must be desired.

The current structure is in the long run harmful for everybody, including those whom it seems to benefit. To be sure, top executives and administrators enjoy successful careers, fat monthly salaries, and sometimes a taste of power. However, these satisfactions are, in most cases, paid for with an intellectual numbness of which they are nonetheless aware; they have scarcely any involvement in the movement of ideas; at the most, they can, by bureaucratic inertia, impede this movement. Because of their premature and absurdly permanent inclusion in the category of the "elite," they become, humanly speaking, victims.

Culture and "intelligence" are as fragile as health. Can one imagine a society where a stay of a few weeks in the hospital would be offered to all young people, after which they would be supposed to benefit for the rest of their lives from the capital of health thus accumulated?

Once the material problems of survival are resolved, society should be organized so as to satisfy the human need for knowledge. By knowledge I mean not the passive absorption of a mass of random and disorganized facts but the process by which a person actively sorts and assimilates this mass and eventually develops questions and answers of his own. This need is unlimited; at present, our responses to it are mere caricatures. Consider, for instance, so-called "mass" tourism, which is supposed to satisfy the need to know other countries and cultures. It is organized in such a way that it merely locks the visitor into his narrow-mindeness and confirms him in his prejudices, while his role with regard to the visited (to use the expression of M. Ki-Zerbo) is that of a "cultural defoliant."

The transformation necessary in this domain concerns society in its entirety, which must become a huge educational system. The demand for leisure is less a demand for a reduction of work than for the opportunity for deliberate, chosen work. The role of the scientist in this change consists above all in becoming modest: in a society where everyone is learning, science becomes a religion without a clergy.

4. *Collective Contempt*

Racism in many forms, from the most brutal to the most surreptitious, rages; we see it every single day. This fact is not new, it is not specific to our society.

However, the form under which this call to reflexes of intolerance is currently being developed is quite specific; it is in the name of the "recent discoveries of modern science," in the name of "models developed by biology" that people claim to justify the classification of people in hierarchical categories.

If scientific advances really led to such conclusions, it would be necessary to take account of them, whatever our personal moral, philosophical, or religious stands on this subject. It so happens that the current content of scientific discourse, particularly in the discipline most relevant to this issue, genetics, says exactly the opposite to what certain people would like to imply that it says. A veritable misinterpretation is necessary in order to found elitist theories on biology, whether the "elite" be constituted by certain individuals within each group or by certain groups.

The current situation calls for a reaction by scientists, whose duty it is to specify and communicate the findings of their various disciplines; racism must be countered, not only with arguments of the heart, but with those of reason. To do this, it is necessary, above all, to be clear, and therefore to carefully define the meaning of the terms being used and, to begin with, that of the word "racism."

To be "racist" is to despise another because of his belonging to a group.

This group can be defined as a function of very diverse criteria: skin color, language, religion, genetic heritage, or cultural heritage; it is more accurate, therefore, to speak of racisms than of racism. Each of these racisms implies that it is possible, on the one hand, to classify people in relatively homogeneous and distinct categories and, on the other, to establish a scale of values between these categories—in other words, to define "races," and then to place them within a hierarchy.

We saw on several occasions in this book that the findings and conclusions of biology allow us to demonstrate both the inanity of the concept of race and the impossibility of founding a hierarchy between individuals or between groups on their genetic heritage. Notwithstanding this, "racism" is a fact.

The natural reaction to physical appearances or behaviors different from our own should, however, be one of wonder. Here again, recounting some specific anecodotes will be more efficacious than a statement of principle.

MAN-MADE PARADISES

In the south of Adrar des Ifoghas, a mountainous massif bordering on the Tanezrouft desert a few hundred kilometers from the Niger, I am spending a long afternoon in the tent of Amatu, *amenukal* of the Touareg kel-Kummer. All the other members of the team that had come to complete the taking of blood samples from this tribe have gone for the day to another campground; for some reason, I stayed on. No wind, a pitiless sun; around the tent, a few thorny shriveled shrubs stuck in the sand; nothing moves; camels and goats, stretched out, wait. Beside me, two old women are playing at moving pegs on a board drawn by their fingers on the sand: one has white pegs, pebbles, the other black ones, camel scat. From time to time, the less old of the two raises the base of the tent here, lowers it there, creating a draft that is pleasant and almost fresh. From a bowl I am slowly drinking the last few drops of camel's milk. My feeling of well-being is infinite. . .

Billiam, one of the last seal hunters of the little village of Tileqilaq, is bringing us up the Kulusuk, a huge fjord on the east coast of Greenland; he guides us between the icebergs, breathtaking masses of ice that have come down from the polar icecap, slowly drifting toward the Atlantic. In the July sun they are melting and taking on fantastic shapes; no imagination is required to see oneself surrounded by fabulous, fixed animals, to imagine oneself in the ruins of some architecturally extravagant forgotten city. Billiam takes us to the entrance of a creek near a stream pouring down from the mountain; he casts out his nets; soon ten, twenty salmon are caught. We cook some of them. It is one o'clock in the morning; the sun, hidden for the past few minutes behind a peak, reappears and brings everything back to life; we are chilled to the bone, so extreme is the cold. However, the taste of the salmon is beyond compare, the sun is making the ice glisten all around us; how could one fare better than this? . . .

This pitiless desert where, for ten months at a stretch, not a single drop of water falls, this region of ice and wind where nothing can grow, are they disinherited lands? No, thanks to humans, they are paradise sometimes. Man-made paradises, since nature made no concessions, except for the incredible beauty of the landscape; it provided only unbearable heat or cold, drought or flooding, the parching harmattan or the piercing blizzard.

We humans were able to invade the planet and create conditions favorable to our survival or, better still, our well-being. Many other species on the planet slowly evolved organisms suitable to their environmental conditions but this would have taken us too long; instead, we were able to adapt our way of being and to transform the environment while keeping what is essential: our capacity for observing our surroundings, while keeping our distance from them.

Robert Gessain, who lived for a long time with the Eskimos of Ammassalik, shows us how, before being assimilated into our civilization, they had conquered death:[13] each baby, by receiving the name of a deceased relative, resurrected the latter; the endless cycle of souls guaranteed the eternity of each individual and that of the group. Still more amazingly, this eternity was extended to the seals, which they had to hunt and kill; the seal's soul was also to be reincarnated in the next baby seal; death was but the gift of its provisional body and, before eating, the Ammassalimiut politely said as a kind of grace, "Thank you, seal, for having given me your meat."

For the Touareg kel-Kummer, death is such an anguishing mystery that even the names of the dead cannot be uttered: each baby receives a name that no one had ever had previously and that will never again be given.[14] Wherever a person dies, he is buried unceremoniously and wordlessly; scarcely does a slight mound mark off the location, and, as long as any trace of it remains, everyone avoids going too close to it.

Is an exchange possible between those who come to terms with death by denying the obvious (but the obvious is often misleading) and those who have banished it from their daily lives by trying to ignore it? If the kel-Kummer and the Ammassalimiut, both ingenious inventors of gestures, of tools, and of explanations of the world, were to meet, how would they react?

It is probable, alas, that they would despise each other, as we despise those who behave differently from us. What therefore is this poison that we distill in reaction to others? How is it that we, who are so well able to observe the world around us, to use it to the best advantage, and to modify it, are so poorly equipped to look at other human beings? At the present time, do we not need, more than anything else, to learn to delight in difference?

COMBATING CONTEMPT

Scattered throughout the planet, we have been conquerors everywhere: how can one imagine a human achievement greater than our ability to smile, totally content, in an igloo inside the Arctic circle and in a tent in the desert? How can we accept the idea of spoiling these victories, our victories, by despising those who have brought them about?

In order to combat this behavior, it is necessary to face up to it, to take it for what it is—a perversion—to study its causes, and to look for remedies. It is therefore a truly scientific attitude that is required, and UNESCO is trying to promote such an attitude; in April 1981, it organized in Athens a meeting of researchers from different disciplines and nations in order to draft "A call to the peoples of the world and to each and every human being," which was adopted unanimously. It states in particular: "Genetic diversity is present much more between individuals belonging to the same population than between the statistical averages of these populations, which does away with any possibility of an objective and stable definition of human races . . . . Participating in science means having a significant share in responsibility for the social future of one's contemporaries. This responsibility implies, in the face of racism, political and ethical choices."

The implementation of this call would by itself alone bring about a sweeping change in the behavior of many societies, certainly of our own.

Our hierarchization reflexes, which lead us to confuse nonequality, that is, difference, with inequality, that is, a relationship of superiority-inferiority, do not only result, as we have described, from the grip of elementary arithmetic on our minds. They are reinforced in our civilization by the role of money. Everything can be situated on a single scale of values: monetary values. This scale is imposed on us because everything in our daily life depends on our salary and on the cost of what we desire.

We are accustomed to categorizing and classifying individuals and groups according to what they are worth, that is, according to the money that they have at their disposal.

Social mechanisms for distributing tasks according to their unpleasantness, for creating groups with limited rights, immigrants, and for housing certain categories of people in certain sections of cities have created undeniable barriers between coexisting groups. Globally satiated, our society is fearful; in times of unrest, it seeks a scapegoat. Though no longer subject to the ancestral anguish of hunger and cold, people remain worried since the portion of wealth granted to them could be reduced; despised by those whose portion is greater, they compensate by despising those whose portion is lesser.

Our society secretes racism. Usually without being aware of it, we secrete this poison that is destroying us; we say, "I am not a racist, but. . ." We pointed out that an editorial writer who, naturally, declares and believes himself to be nonracist could recently proclaim the necessity for distinguishing "the s____ from the diamond" in humanity.

We cannot prevent those who want, or need, to make this distinction. However, when they claim to do it in the name of science, it is the duty of scientists to denounce the swindle. For biologists, this duty takes precedence over the more generally recognized one of contributing to the advancement of knowledge, because they can teach us a constructive way of looking at people who are different from us.

5. *Another Way of Looking at Man*

Like all living beings, a child receives at the outset a heritage of genetic information containing the codes which will enable his organism to construct, develop, and struggle to maintain himself. The complex substances of which he will be constituted, the subtle regulatory mechanisms that will stabilize him, the internal clocks that will determine the timing of the successive phases of his development and then of his aging are irrevocably defined by the collection of genes received half from his father, half from his mother.

But these genes, in isolation, are mute; it is only because of input from the environment that they can be expressed. During the nine months of fetal life, this environment is the maternal womb, after which its scope broadens to include everything and everybody that provides nourishment, energy, information, and affection.

The living beings that appeared a few billion years ago, our distant ancestors, also needed this double input: genes and environment. However, they were mere "genetic automats," tirelessly repeating behaviors defined by their genes. Gradually, a certain flexibility appeared, permitting the transfer of the procreator's experience to the progeny; a learning process assured the transmission of information not encoded in the genes.

This learning process lasts an especially long time in our species, but it is not specific to us. Certain birds have to learn the song of their species: it is not inscribed on their genes. For primates, our closest relatives, the learning process lasts several years.

On the contrary, we are the only living beings to have succeeded in instituting a third source of information: as well as his genes and the environment in which he finds himself, a child, thanks to language and writing, has at his command a memory, external to himself and to others, where all human experience is inscribed. The child who is in the process of developing is rich not only with the biological recipes provided by his genes, not only with substances provided by his environment and the behavior taught by his entourage, but potentially with the treasures accumulated by the oral traditions or stored in all the libraries of the world.

At the junction of these three streams, man was able, thanks to their interaction, to benefit from an extraordinary privilege: that of creating himself. If we represent this process of realization by a diagram (see figure 8), we need to add to the three arrows symbolizing the contributions of genes, environment, and society a fourth arrow taking off

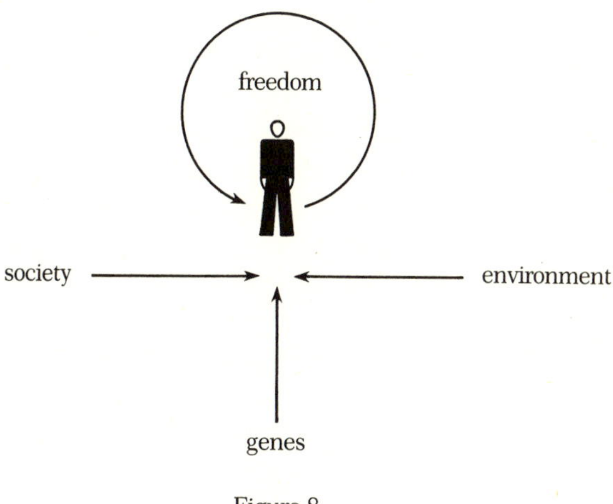

Figure 8

from the individual and returning to him. This arrow symbolizes the capacity for autostructuring resulting from our complexity.

This notion of complexity is being currently emphasized by biologists and also by physicists.[15] The most rudimentary living organism shows such complexity, both in structure and in functioning, that it is impossible to describe it adequately if we deal only with its components. Properties specific to the whole must be taken into account; they concern in particular the power possessed by a sufficiently complex material structure to benefit from a disturbance that it undergoes in order to autostructure, enrich, and prepare itself to react still more efficaciously to future disturbances. The mishaps, the aggressions, the "noise," instead of being destructive, can become sources of enrichment.

Let us attempt to more clearly define this veritable slap-in-the-face given by current scientific thinking to the classical theories, which have for so long been the background for our understanding of the universe.

TIME AS MOWER AND TIME AS SOWER

The effect of the passage of time on the structure of material entities is defined by means of a central concept developed in the nineteenth century, that of "entropy." By this term, the physicist denotes the seemingly irreversible decline brought about by aging: every organization is gradually destroyed by use, every piece of information is partially deformed during its transmission, every arrangement is gradually disturbed by its own inner dynamic. Inexorably, disorder sets in; in scholarly terms, "entropy increases."

Duration, when incorporated into our vision of the real world, is thus associated with wear and tear, decline, and destruction; it is represented by the image of a skeleton, armed with a scythe, wreaking death upon everything along its path. This vision was given scientific formulation by Sadi Carnot who, in 1824, developed the "second principle of thermodynamics." The first, called "the principle of energy conservation," states that the total energy of an isolated system remains constant; this is a quantitative finding. The second concerns the quality of this energy; Carnot shows that it can only deteriorate. With each transformation, a less "high-quality" energy replaces the former energy. Ultimately, at the end of this process, all structure has disappeared, all organization has collapsed, every link has been broken, every object has been diluted; the universe is no longer anything other than an undifferentiated grayness, more sinister than nothingness.

However, this result, which people are willing to generalize without giving it too much thought, is valid, strictly speaking, only for isolated systems. In reality, only one "system," the universe as a whole, is strictly isolated. Does this really interest us when we are incapable of defining it?

In reality, all the systems that interest us, that are really part of our actual experience, are parts of the universe; they are not isolated; they are crossed by a flow of energy and subject, on their boundaries, to perturbations; they constitute, in the terminology of Ilya Prigogine, "dissipative structures." For them, the second principle of thermodynamics takes quite a different form: the accidental fluctuations that they undergo may provoke in them new and more complex structures, richer in reactive possibilities than the previous structures.

Time is not therefore necessarily destructive—it provides opportunities for creation. It does not actually create, but it opens the door to spontaneous creations. Its image becomes that of time as creator, as sower. The traditional image of a skeleton armed with a scythe should be replaced by that of a farmer walking along the furrow, holding a bag of seeds that he scatters around him and that promise new life.

The history of the living world is an obvious illustration of this creative role of time. The first whisperings of life, molecules capable of autoreproduction, permitted the appearance of increasingly complex structures, of organizations capable of increasingly differentiated functions, of beings developing new powers. The strangest case is undoubtedly that of our own species, which, as Boris Vian put it, found itself endowed with an excess "of rather soft phosphorus, a brain that enabled us to foresee ourselves without life." Thanks to this brain, we have the power to destroy ourselves, both individually and collectively, and also to autostructure and invent ourselves.

The sweeping changes required to go from blue seaweed to man have all involved enormous risks, but luckily these changes have introduced new possibilities. How many failures, drownings, and catastrophes happened in the course of evolution which sacrificed thousands of species to culminate, by divergent paths, in the few million species that subsist, one of which is *Homo sapiens*!

The life span of each of us is less than a century long, but it is as eventful as the history of life throughout its three billion years. It too involves leaps and sweeping changes; the end result was not inscribed in our initial state; the chance of successive transformations constitutes a hold on ourselves, which we can grasp.

AUTOSTRUCTURING AND FREEDOM

It is therefore from observing the behavior of the real world that I was led to add to my diagram representing the input and the influences thanks to which a man makes himself a fourth arrow that rebounds back on itself. It symbolizes his capacity for using, to structure himself, the random perturbations inflicted on him by the outside world; what could be destructive becomes creative.

Can this last arrow not be seen as a symbol of freedom? The individual is not only a product of his genes, his environment, and of the society to which he belongs; he is also a subject who, in part, creates himself.

For this to be possible, input from the other three sources must have been sufficient to allow him to achieve sufficient complexity, and, above all, outside influences, those of family or of society, must not have intervened to prevent this autostructuring from developing. A social system is, it seems to me, "fascist" in so far as it denies this fourth arrow and considers each being to be the result of outside influences alone. Even if he receives all the attention necessary to give him a vigorous body and a well-filled mind, he is, in the last resort, only a fabricated object if he is not allowed to participate in the process of his own creation, or if he is conditioned in such a way as to avoid this self-creation.

Every effort should, on the contrary, strive to provoke this taking of responsibility by each individual for his own life. On this path, scientific activity, as it is practiced in reality and not as it is presented by those who use it to camouflage their ideology, can play a decisive and truly liberating role. It is science that trains us to ask better questions, and therefore to "be" more fully; because to be, is it not first of all to ask questions of ourselves?

Notes

1. Science and Us

1. I. Prigogine and I. Stengers, *La Nouvelle Alliance* (Paris: Gallimard, 1979).

2. The "cardinal" of a set is the number of its elements.

3. Claudel made a distinction between "connaissance," the acquisition of knowledge, and "co-naissance" (literally "being born with"), which emphasizes the process rather than the products of knowledge.

4. Translator's note: *Topaze,* a play published in 1930, depicts a naive and honest teacher who gradually becomes versed in the ways of the world.

2. Number Pitfalls

1. These lines are from Baudelaire's famous poem "L'Invitation au voyage."

2. For instance, I. Prigogine and I. Stengers in *La Nouvelle Alliance* (Paris: Gallimard, 1979).

3. In effect, 1/8 of the population lives in environment I and has the a_1a_1 genotype; 2/8 lives in environment I with the a_1a_2 genotype, etc. Hence

$$m = \tfrac{1}{8}\,100 + \tfrac{2}{8}\,140 + \tfrac{1}{8}\,60 + \tfrac{1}{8}\,70 + \tfrac{2}{8}\,90 + \tfrac{1}{8}\,110 = 100.$$

4. Similarly, by definition, the variance is

$$V = \tfrac{1}{8}\,(100-100)^2 + \tfrac{2}{8}\,(140-100)^2 + \tfrac{1}{8}\,(60-100)^2 + \tfrac{1}{8}\,(70-100)^2 + \tfrac{2}{8}\,(90-100)^2 + \tfrac{1}{8}\,(110-100)^2 = 750.$$

5. In effect, in environment I the average is

$m_{\mathrm{I}} = \tfrac{1}{4}\,100 + \tfrac{1}{2}\,140 + \tfrac{1}{4}\,60 = 110$, and the variance is

$$V_{\mathrm{I}} = \tfrac{1}{4}\,(100-110)^2 + \tfrac{1}{2}\,(140-110)^2 + \tfrac{1}{4}\,(60-110)^2 = 1{,}100.$$

3. *The Pitfalls of Classification*

1. S. Karlin, R. Kenett, and B. Bonné-Tamir, "Analysis of Biochemical Genetic Data on Jewish Populations," *American Journal of Human Genetics* (1979), 31:341–365.

2. M. Greenacre and L. Degos, "Correspondence Analysis of HLA Genes Frequency Data from 124 Population Samples," *American Journal of Human Genetics* (1977), 29:60–75.

3. See A. Jacquard, *In Praise of Difference* (New York: Columbia University Press, 1984), p. 84.

4. Cited by A. Langaney, "Diversité et histoire humaines," *Population* (1979), 34:985–1005.

5. The data presented here are taken from a study by Alcantara, Bordier, and Obadia, in Professor J.-P. Benzécri's team at l'Université Pierre-et-Marie-Curie (Paris VI).

6. A more detailed analysis shows that the point "Ducatel" is close to the point representing "blank" votes: voting for this candidate, whose chances were nil and whose political stand was poorly defined, was scarcely different from abstaining.

4. *The Pitfall of Words*

1. For the simple reason that the conditions for the test vary constantly; the composition of the urn is changed at each draw.

2. J. Stewart, W. Debray, and V. Caillard, "Schizophrenia: The Testing of Genetic Models by Pedigree Analysis," *American Journal of Human Genetics* (1980) 32:55–63.

5. *Biology And Education: Intelligence, Its Support, and Its Development*

1. J.-P. Changeux and A. Danchin, "Apprendre par stabilisation sélective de synapses en cours de développement," in E. Morin and M. Piattelli-Palmarini, *L'Unité de l'homme*, vol. 2, *Le Cerveau humain* (Paris: Seuil), coll. Points-Sciences humaines (1974), 92:58–84.

2. A. Danchin, *Ordre et dynamique du vivant* (Paris: Seuil, 1978).

3. T. Tsunoda, *The Mother Tongue and Right-Left Dominance in the Human Central Auditory System* (Paris: UNESCO, 1981).

4. H. Atlan, *L'Organisation biologique et la théorie de l'information* (Paris: Hermann, 1972).

5. F. Jacob, "Sexualité et diversité humaine," *Le Monde*, February 9, 1979.

6. The *Robert* is one of the most detailed and respected dictionaries of the French language.

7. P. Dague, *La Mesure de l'intelligence*, Colloque du MURS, Paris, 1977.

8. M. Carlier, "Pour un bon usage de la notion de QI," in *L'Intelligence est-elle héréditaire?* (Paris: ESF, 1981).

9. P. Debray-Ritzen, *Lettre ouverte aux parents des petits écoliers* (Paris: Albin Michel, 1978).

10. N. Morton, "Effect of Inbreeding in IQ and Mental Retardation," *Proceedings of the National Academy of Sciences* (August 1978), no. 8.

11. H. Eysenck, *L'Inégalité de l'homme* (Paris: Copernic, 1977).

12. D. Dorfman, "The Cyril Burt Question: New Findings," *Science* (29 September 1978), 201:1177–1186.

13. Michel Schiff, et al., *Enfants de travailleurs manuels adoptés par des cadres* (Paris: INED/PUF, 1981).

14. C. Bert, "Les surdoués," *Le Monde de l'Education* (November 1978) no. 44, pp. 58–65.

6. *Biology and Social Organization: Sociobiology*

1. M. Sahlins, *Critique de la sociobiologie* (Paris: Gallimard, 1980).
2. A. Langaney, *Le Sexe et l'innovation* (Paris: Seuil, 1979).
3. M. Rouzé, "Us et abus de la biologie," *AFIS* (1980), vol. 93.

7. *The Evolution of the Living World: Facts and Explanatory Models*

1. A fact now accepted without restriction, but this is recent: the teaching of the theory of evolution was forbidden by law in the state of Tennessee until 1968 when this law was declared unconstitutional by the Supreme Court of the United States.

2. M. Nei, *Molecular Population Genetics and Evolution* (New York: North-Holland, 1975).

3. Remember that a locus is the position occupied on a chromosome by the genes governing a given elementary trait.

4. E. Nevo, "Genetic Variation in Natural Populations: Patterns and Theory," *Theoretical Population Biology* (1978) 13:121–178.

New Departures

1. J.-M. Legay, *Qui a peur de la science?* (Paris: Éditions sociales, 1981).

2. See, for example, M. Agi, *De l'idée d'universalité comme fondatrice du concept des droits de l'homme* (Antibes: Alpazur, 1980).

3. Author, notably, of *Utopies réalisables* (Paris: UGE), coll. "10/18," 1975; *Comment vivre entre les autres sans être chef et sans être esclave* (Paris: Pauvert, 1974); *L'Architecture de survie* (Paris: Casterman, 1978).

4. This text was written before the animators of Radio-LCA were fired and "order was restored."

5. R. J. Herrnstein, *IQ in the Meritocracy* (Boston: Little, Brown, 1973).

6. P. Debray-Ritzen, *Lettre ouverte aux parents des petits écoliers* (Paris: Albin Michel, 1978).

7. M. Schiff, "L'échec scholaire n'est pas inscrit dans les chromosomes," *Psychologie* (December 1980), 131:51–56.

8. Cahiers "Travaux et documents" de l'INED: *Niveau intellectuel des enfants d'âge scolaire*, 1950, no. 13; *Le Niveau intellectuel des enfants d'âge scolaire: La détermination des aptitudes, l'influence des facteurs constitutionnels, familiaux*

et sociaux, 1954, no. 23; *Enquête nationale sur le niveau intellectuel des enfants d'âge scolaire,* 1969, no. 54; *Enquête nationale sur le niveau intellectual des enfants d'âge scolaire,* 1978, no. 83.

9. J. Capelle, "Les CES ont-ils échoué?" *Le Monde,* February 2, 1977.

10. M. Schiff, A paper given at the sixth Human Genetics Congress, Jerusalem, 1981.

11. *Ibid.*

12. I. Illich, Deschooling Society (New York: Harper and Row, 1983).

13. R. Gessain, *Ammassalik ou la civilisation obligatoire* (Paris: Flammarion, 1970).

14. A. Chaventre, *Les Touareg kel-Kummer* (Paris: INED/PUF, 1982).

15. For instance, H. Atlan, *Entre le cristal et la fumée* (Paris: Seuil, 1979); or I. Prigogine and I. Stengers, *La Nouvelle Alliance* (Paris: Gallimard, 1979).

DATE			